NANOTECHNOLOGY WORKFORCE PROTECTION

STRATEGIC PLAN AND SAFE PRACTICES

NANOTECHNOLOGY SCIENCE AND TECHNOLOGY

Additional books in this series can be found on Nova's website under the Series tab.

Additional e-books in this series can be found on Nova's website under the e-book tab.

NANOTECHNOLOGY SCIENCE AND TECHNOLOGY

NANOTECHNOLOGY WORKFORCE PROTECTION

STRATEGIC PLAN AND SAFE PRACTICES

SUZANNE G. RICE
EDITOR

nova publishers
New York

For permission to use material from this book please contact us:
Telephone 631-231-7269; Fax 631-231-8175
Web Site: http://www.novapublishers.com

Library of Congress Cataloging-in-Publication Data

ISBN: 978-1-63117-641-8

Published by Nova Science Publishers, Inc. † New York

CONTENTS

PREFACE

Nanotechnology—the manipulation of matter on a near-atomic scale to produce new materials and devices—has the ability to transform many industries, from medicine to manufacturing, and the products they produce. Nanomaterials may present new challenges to understanding, predicting, and managing potential health risks to workers. Many knowledge gaps still remain on how to work safely with all of these materials. Through strategic planning, research, partnering with stakeholders, and making information widely available, the National Institute for Occupational Safety and Health (NIOSH) is working to continue to provide national and world leadership in providing solutions that will prevent work-related illness and injury. NIOSH hopes to raise awareness of the occupational safety and health practices that should be followed during the synthesis, characterization, and experimentation with engineered nanomaterials in a laboratory setting. This book contains recommendations on engineering controls and safe practices for handling engineered nanomaterials in laboratories and some pilot scale operations. The book was designed to be used in tandem with well-established practices and the laboratory's chemical hygiene plan. As our knowledge of nanotechnology increases, so too will our efforts to provide additional guidance materials for working safely with engineered nanomaterials.

Chapter 1 – The National Institute for Occupational Safety and Health (NIOSH) is pleased to present *Protecting the Nanotechnology Workforce: NIOSH Nanotechnology Research and Guidance Strategic Plan, 2013–2016.* This plan updates the November 2009 strategic plan with knowledge gained from results of ongoing research, as described in the 2012 report *Filling the Knowledge Gaps for Safe Nanotechnology in the Workplace: A Progress Report from the NIOSH Nanotechnology Research Center, 2004–2011.* The

NIOSH Nanotechnology Research Program follows a comprehensive plan that is managed as a matrix structure across NIOSH and supports multiple sectors in the National Occupational Research Agenda (NORA).

Nanotechnology provides many opportunities for advancing the economic value and impact of new U.S. technologies and products as it expands into every industrial sector. Today, nanomaterials are found in hundreds of products, ranging from cosmetics, to clothing, to industrial and biomedical applications. The potential benefits of nanotechnology are huge, and these benefits should be realized by society. However, there is ongoing concern that the full potential of the societal benefits may not be realized if research efforts are not undertaken to determine how to best manage and control the potential occupational safety and health hazards associated with the handling of these nanomaterials.

Chapter 2 – The National Institute for Occupational Safety and Health (NIOSH) is pleased to present *General Safe Practices for Working with Engineered Nanomaterials in Research Laboratories.* Engineered nanomaterial applications are rapidly expanding throughout the United States and worldwide. The research community is at the front line of creating these new nanomaterials, testing their usefulness in a variety of applications, and determining their toxicological and environmental impacts.

With the publication of this document, NIOSH hopes to raise awareness of the occupational safety and health practices that should be followed during the synthesis, characterization, and experimentation with engineered nanomaterials in a laboratory setting. The document contains recommendations on engineering controls and safe practices for handling engineered nanomaterials in laboratories and some pilot scale operations. This guidance was designed to be used in tandem with well-established practices and the laboratory's chemical hygiene plan. As our knowledge of nanotechnology increases, so too will our efforts to provide additional guidance materials for working safely with engineered nanomaterials.

In: Nanotechnology Workforce Protection
Editor: Suzanne G. Rice
ISBN: 978-1-63117-641-8

Chapter 1

PROTECTING THE NANOTECHNOLOGY WORKFORCE: NIOSH NANOTECHNOLOGY RESEARCH AND GUIDANCE STRATEGIC PLAN, 2013–2016*

National Institute for Occupational Safety and Health

FOREWORD

The National Institute for Occupational Safety and Health (NIOSH) is pleased to present *Protecting the Nanotechnology Workforce: NIOSH Nanotechnology Research and Guidance Strategic Plan, 2013–2016*. This plan updates the November 2009 strategic plan with knowledge gained from results of ongoing research, as described in the 2012 report *Filling the Knowledge Gaps for Safe Nanotechnology in the Workplace: A Progress Report from the NIOSH Nanotechnology Research Center, 2004–2011*. The NIOSH Nanotechnology Research Program follows a comprehensive plan that is managed as a matrix structure across NIOSH and supports multiple sectors in the National Occupational Research Agenda (NORA).

* This is an edited, reformatted and augmented version of DHHS (NIOSH) Publication No. 2014–106, dated December 2013; released by U.S. Department of Health and Human Services, Centers for Disease Control and Prevention, National Institute for Occupational Safety and Health.

Nanotechnology provides many opportunities for advancing the economic value and impact of new U.S. technologies and products as it expands into every industrial sector. Today, nanomaterials are found in hundreds of products, ranging from cosmetics, to clothing, to industrial and biomedical applications. The potential benefits of nanotechnology are huge, and these benefits should be realized by society. However, there is ongoing concern that the full potential of the societal benefits may not be realized if research efforts are not undertaken to determine how to best manage and control the potential occupational safety and health hazards associated with the handling of these nanomaterials.

The research conducted over the past 8 years has proven that NIOSH is a global leader in promoting the responsible development of nanotechnology. NIOSH has built business partnerships, established itself as a key player in nanotoxicology, published precautionary guidance (*Approaches to Safe Nanotechnology: Managing the Safety and Health Concerns Associated with Engineered Nanomaterials*), and issued recommended exposure limits for nanoscale titanium dioxide and for carbon nanotubes and nanofibers.

This *NIOSH Nanotechnology Research and Guidance Strategic Plan* is the roadmap being used to advance basic understanding of the toxicology and workplace exposures involved so that appropriate risk management practices can be implemented during discovery, development, and commercialization of engineered nanomaterials. NIOSH will strive to remain at the forefront of developing guidance that supports and promotes the safe and responsible development of such a promising technology.

John Howard, M.D.
Director, National Institute for Occupational
Safety and Health
Centers for Disease Control and Prevention

EXECUTIVE SUMMARY

Nanotechnology—the manipulation of matter on a near-atomic scale to produce new materials and devices—has the ability to transform many industries, from medicine to manufacturing, and the products they produce. By 2020, the National Science Foundation estimates, nanotechnology will have a $3 trillion impact on the global economy and employ 6 million workers in the manufacture of nanomaterial-based products, of which 2 million may be manufactured in the United States [NSF 2011]. Nanomaterials may present new challenges to understanding, predicting, and managing potential health risks to workers.

Many knowledge gaps still remain on how to work safely with all of these materials. Through strategic planning, research, partnering with stakeholders, and making information widely available, the National Institute for Occupational Safety and Health (NIOSH) is working to continue to provide national and world leadership in providing solutions that will prevent work-related illness and injury.

Nanotechnology and NIOSH Research

Nanotechnology and the commercialization of products and devices containing engineered nanomaterials could help address critical global problems concerning energy, transportation, pollution, health, and food. The potential benefits of nanotechnology are huge, and these benefits should be realized. Nonetheless, there is concern that the full potential of the societal benefits may not be realized if cautions about the adverse human health effects are not heeded and concerns are not honored. Timely, targeted research is needed to define hazards, exposures, and risks and to provide guidance for safe handling of nanomaterials. A concerted effort by industry, academia, labor, the professions, and government is needed to identify and address the knowledge gaps in a transparent and credible process that coincides with development of this new technology. NIOSH is playing an active part in this process by supporting the development of a broad spectrum of research and prevention strategies for health and safety hazards related to nanotechnology. In a series of reports [NIOSH 2007, 2009a, 2012a], NIOSH has summarized its progress in conducting nanotechnology research and recommending risk management strategies (see http://www.cdc.gov/niosh/topics/nanotech/). NIOSH investigators have identified adverse health effects in animals exposed to various engineered nanomaterials; assessed exposure of workers; initiated epidemiologic research; and provided guidance on control technologies and medical surveillance. There are many questions still to be answered. A vast number of potential new nanomaterials are possible and could result in a seemingly limitless combination of physicochemical factors. There is need for an expeditious approach to controlling exposure to the large number of nanomaterials in science and commerce now and in the future. Moreover, the advanced nanomaterials under development are likely to have additional potentially hazardous characteristics that will need to be addressed [Murashov et al. 2012].

NIOSH Nanotechnology Research Center (NTRC)

The NIOSH Nanotechnology Research Center (NTRC) was established in 2004 to coordinate nanotechnology research across the institute. Ten critical areas of research were identified, each having at least one key scientist serving as a coordinator. The NTRC and its steering committee of critical area coordinators are responsible for developing and guiding NIOSH scientific and organizational plans in nanotechnology health and safety research.

Strategic Plan

The development of nanotechnology has reached a point where it is being widely applied, and numerous nanomaterials and nano-enabled products are in commerce. Nanotechnology has the potential to provide great benefit to society, but it must be developed responsibly. This responsibility involves addressing any adverse human and environmental impacts of the technology associated with engineered nanomaterials (ENMs). Workers are among the first people in any society to be exposed to the potential health hazards caused by the products of new technology, and their exposure to any new material is often greater than for the general population. Therefore, worker safety and health can be seen as the core of responsible development (Figure 1).

Through its strategic plan for fiscal years (FY) 2013–2016, NIOSH will marshal its resources and partner with others efficiently and effectively to advance efforts to protect the nanotechnology workforce. With the input of a broad range of stakeholders in government, academia, and the private sector, NIOSH will continue to operate under a strategic plan for nanotechnology research and guidance. The most recent previous version was published in November 2009 and included research plans through FY 2012 (see http://www.cdc.gov/niosh/docs/2010-105/).

This document presents the NTRC Strategic Plan for FY2013–FY2016. The strategic plan also highlights how the critical research and guidance efforts of NIOSH align with and support the comprehensive Environmental, Health, and Safety Research Strategy needs of the National Nanotechnology Initiative. For the period FY2013–FY2016, NIOSH will continue to fill information and knowledge gaps that address the five NIOSH NTRC strategic goals:

1. Increase understanding of new hazards and related health risks to nanomaterial workers.
2. Expand understanding of the initial hazard findings of engineered nanomaterials.
3. Support the creation of guidance materials to inform nanomaterial workers, employers, health professionals, regulatory agencies, and decision-makers about hazards, risks, and risk management approaches.
4. Support epidemiologic studies for nanomaterial workers, including medical, cross-sectional, prospective cohort, and exposure studies.
5. Assess and promote national and international adherence with risk management guidance.

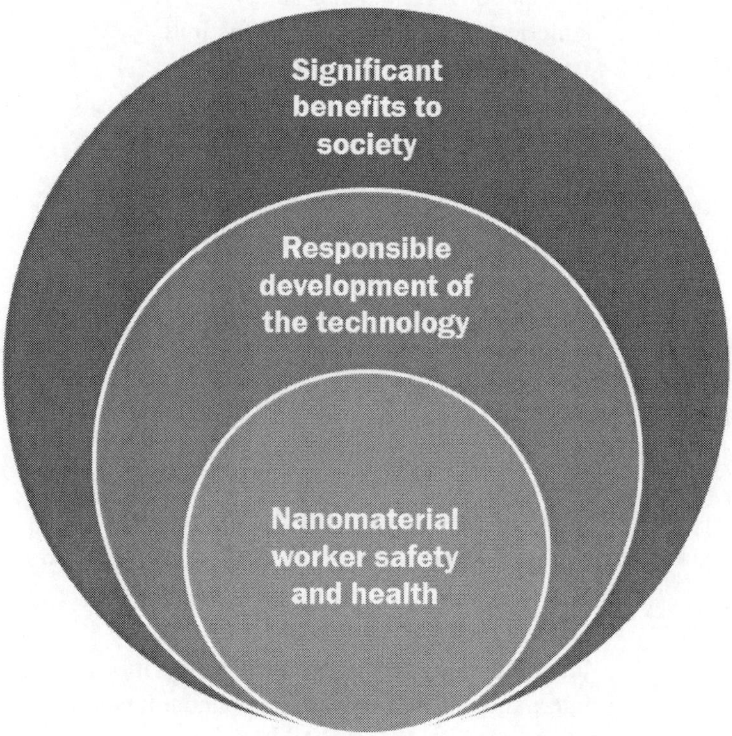

Figure 1. The core of responsible development of nanotechnology.

To address these strategic goals and promote the responsible development of engineered nanomaterials, the strategic plan will expand research activities in 10 NTRC critical areas: toxicity and internal dose; measurement methods; exposure assessment; epidemiology and surveillance; risk assessment; engineering controls and personal protective equipment (PPE); fire and explosion safety; recommendations and guidance; global collaborations; and applications.

Abbreviations

AFL-CIO	American Federation of Labor and Congress of Industrial Organizations
AIHA	American Industrial Hygiene Association
ANSES	The French Agency for Food, Environmental and Occupational Health and Safety
ANSI	American National Standards Institute
ASSE	American Society of Safety Engineers
ASTM	American Society for Testing and Materials (currently ASTM International)
CNT	carbon nanotube
CPSC	Consumer Product Safety Commission
DHHS	Department of Health and Human Services
DOT	Department of Transportation
EHS	environment, health and safety
ENM	engineered nanomaterial
EPA	U.S. Environmental Protection Agency
EU	European Union
FMSH	Federal Mine Safety and Health
FY	fiscal year
ICOH	International Commission on Occupational Health
ICON	International Council on Nanotechnology
IEC	International Electrotechnical Commission
ILO	International Labour Organization
ISEA	International Safety Equipment Association
ISO	International Organization for Standardization
MSHA	Mine Safety and Health Administration
NanoBCA	Nanobusiness Commercialization Association
NCER	National Center for Environmental Research

NEHI	Nanotechnology Environmental and Health Implications
NFPA	National Fire Protection Association
NIEHS	National Institute of Environmental Health Sciences
NIH	National Institutes of Health
NIOSH	National Institute for Occupational Safety and Health
NIST	National Institute of Standards and Technology
NNI	National Nanotechnology Initiative
NORA	National Occupational Research Agenda
NSC	National Safety Council
NSF	National Science Foundation
NTRC	Nanotechnology Research Center
OECD	Organization for Economic Cooperation and Development
OEP	Office of Extramural Programs
OSH	occupational safety and health
OSHA	Occupational Safety and Health Administration
PPE	personal protective equipment
PtD	Prevention through Design
r2p	Research to Practice
TNO	The Netherlands Organization for Applied Scientific Research
UN	United Nations
WHO	World Health Organization

1. INTRODUCTION

1.1. Background

Nanotechnology is a system of innovative methods to control and manipulate matter at near-atomic scale to produce new materials, structures, and devices. Nano-objects are nanomaterials that have at least one dimension less than 100 nanometers [ISO 2008]. Nanoparticles are a specific class or subset of these nano-objects, having three dimensions that are less than 100 nanometers. Nanoparticles exhibit unique properties because of their nanoscale dimensions. Nanotechnology offers the potential for tremendous improvement and advances in the development of commercial products that may benefit society, such as integrated sensors, semiconductors, medical imaging, drug delivery systems, structural materials, sunscreens, cosmetics, and coatings. Nanotechnology is one of the most enabling technologies across the world. By 2020, the global market for nanotechnology-related products is predicted to reach $3 trillion and employ 2 million workers in the United States alone [NSF 2011]. A review of the 2013 version of the Nanowerk nanomaterials database (http://www.nanowerk.com/phpscripts/n_dbsearch.php) revealed more than 3,000 commercially available nanomaterials.

However, the properties of engineered nanoparticles (e.g., size, surface area, reactivity) that yield many improvements in commercial products may also pose health risks. Increasing numbers of workers are potentially exposed to nanomaterials in research laboratories, start-up companies, production facilities, and operations where nanomaterials are processed, used, disposed, or recycled. The challenge is to determine whether the nature of intentionally produced (engineered) nanostructured materials and devices presents new occupational safety and health risks. At the same time, there is a need to address how the benefits of nanotechnology can be realized while the risks are proactively minimized.

Efforts across multiple federal agencies and the private and academic sectors are fostering the development and use of nanotechnology. In 2001, the President's Council of Advisors on Science and Technology collaborated with the interagency National Science and Technology Council to create the National Nanotechnology Initiative (NNI) [NNI 2001]. This initiative supports basic and applied research in nanotechnology to create new nanomaterials and to disseminate new technical capabilities to industry. The purpose of the NNI is to facilitate scientific breakthroughs and maintain U.S. competitiveness in nanoscience. A stated goal of this interagency program is to ensure that nanotechnology research leads to the responsible development of beneficial applications by giving high priority to research on human health, environmental issues, and societal implications related to nanotechnology.

1.2. Mission of NIOSH

In the Occupational Safety and Health Act of 1970 (OSH Act, Public Law 91-596) and the Federal Mine Safety and Health Act of 1977 (FMSH Act, Public Law 95-164), Congress declared that the intent of these acts was to ensure, insofar as possible, safe and healthful working conditions for every working man and woman, to preserve our human resources. In these acts, NIOSH is given the responsibility for recommending occupational safety and health standards and defining exposure levels that are safe for various periods of employment. These include (but are not limited to) the exposures at which no worker will suffer diminished health, functional capacity, or life expectancy as a result of his or her work experience. By means of criteria documents and other publications, NIOSH communicates these recommended standards to the Occupational Safety and Health Administration (OSHA), the Mine Safety and Health Administration (MSHA), and others in the occupational safety and health community.

Under the OSH Act, NIOSH is charged with conducting "research, experiments, and demonstrations relating to occupational safety and health" and with developing "innovative methods, techniques, and approaches for dealing with [those] problems." The act specifies target areas of research that include identifying criteria for setting worker exposure standards and exploring problems created by new technology in the workplace. In an amendment to the act, NIOSH was given responsibility for conducting training and education "to provide an adequate supply of qualified personnel to carry out the purposes of the Act" and for assisting employers and workers with applying methods to prevent occupational injuries and illness (Section 21 of the OSH Act).

NIOSH has over 40 years of experience in conducting research and formulating recommendations for occupational safety and health. During this period, NIOSH has developed considerable expertise in measuring, characterizing, and evaluating new processes and new materials by conducting quantitative exposure assessments and evaluating health effects. NIOSH also has expertise in developing exposure control systems and prevention strategies as well as experience in conducting risk assessments and recommending effective risk management practices.

In 2003, NIOSH became a member of the Nanoscale Science, Engineering, and Technology (NSET) Subcommittee of the National Science and Technology Council. As a member, NIOSH participates in (1) identifying critical issues related to possible hazards of nanomaterials, (2) protecting worker safety and health in this emerging technology, and (3) developing a strategic plan to address such issues and recommend prevention strategies for the safe handling and use of nanomaterials.

In 2004, NIOSH created the Nanotechnology Research Center (NTRC) to identify critical issues, create a strategic plan for investigating these issues, coordinate the NIOSH research effort, develop research partnerships, and disseminate information gained. The NTRC comprises nanotechnology-related activities and projects supported by approximately 50 scientists in various NIOSH divisions and laboratories. Through the NTRC, NIOSH has identified 10 critical research areas for nanotechnology research and communication.

These 10 critical research areas, updated for FY2013– FY2016, are toxicity and internal dose; measurement methods; exposure assessment; epidemiology and surveillance; risk assessment; engineering controls and personal protective equipment (PPE); fire and explosion safety; recommendations and guidance; global collaborations; and applications. By conducting a complete plan of research in these critical areas in a coordinated, concurrent manner, NIOSH is comprehensively addressing the information and knowledge gaps necessary to protect workers and responsibly move nanotechnology forward so that its far-reaching benefits may be realized.

The NIOSH NTRC is working strategically to fill nanomaterial occupational safety and health knowledge gaps through active intramural and extramural research programs and collaborations. Extramural research is carried out through the NIOSH Office of Extramural Programs (OEP), in which nanotechnology research (through R01, K01, R03, R21, and R43/44 grant mechanisms) is funded to increase the knowledge of nanotechnology and engineered nanomaterials (ENMs) as they relate to occupational safety and health. Research areas supported by the NIOSH OEP include emission and exposure assessment methods for nanoparticles in the workplace, toxicology of ENMs, and use of nanotechnology for the development of sensors. NIOSH is committed to conducting and supporting studies that will improve scientists' abilities to identify potential occupational health effects of nanomaterials. NIOSH will facilitate the translation of those findings into effective workplace practices.

NIOSH reported on the progress of the NTRC in *Filling the Knowledge Gaps for Safe Nanotechnology in the Workplace* (Figure 2). This report presents the program accomplishments of the NTRC from its inception in 2004 through 2011. The NIOSH NTRC continues to be part of the U.S. leadership on the International Organization for Standardization (ISO) TC 229 Nanotechnology Working Group on Health, Safety, and the Environment and continues to work with the World Health Organization (WHO) collaborating centers on global projects of information dissemination and communication. Through these collaborations, the NIOSH NTRC assists in developing risk communications on the safe handling of nanomaterials.

1.3. NIOSH Logic Model

NIOSH receives input on identifying and addressing occupational safety and health problems through a logic model (i.e., a process) in which the seriousness of the problem or hazard is evaluated, the type and level of research needed are determined, and a plan and process for communicating the research outcomes are formulated. The overall NIOSH logic model (Figure 3) has a conventional horseshoe shape, with the operational upper branch proceeding from inputs to mission relevant outcomes and the strategic lower branch supporting those operations through a process of measurable goals and management objectives. The two branches are correlated and are subject to external factors.

The NIOSH operational model (Figure 4) for conducting research adheres to the logic model in the acquisition and analysis of inputs from customers/stakeholders (*production inputs*) and internal/external research capabilities (*planning inputs*) to determine and prioritize research.

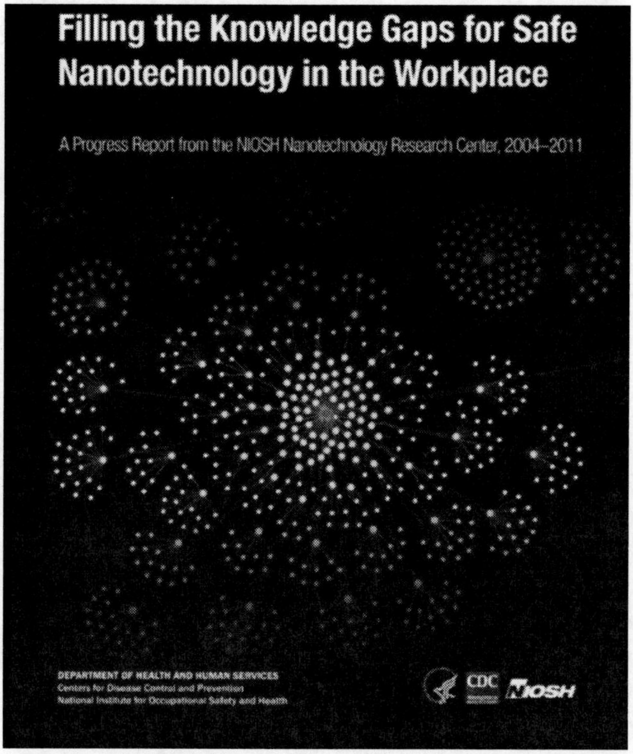

Figure 2. The cover of DHHS (NIOSH) Publication No. 2013–101, the progress report of the NTRC.

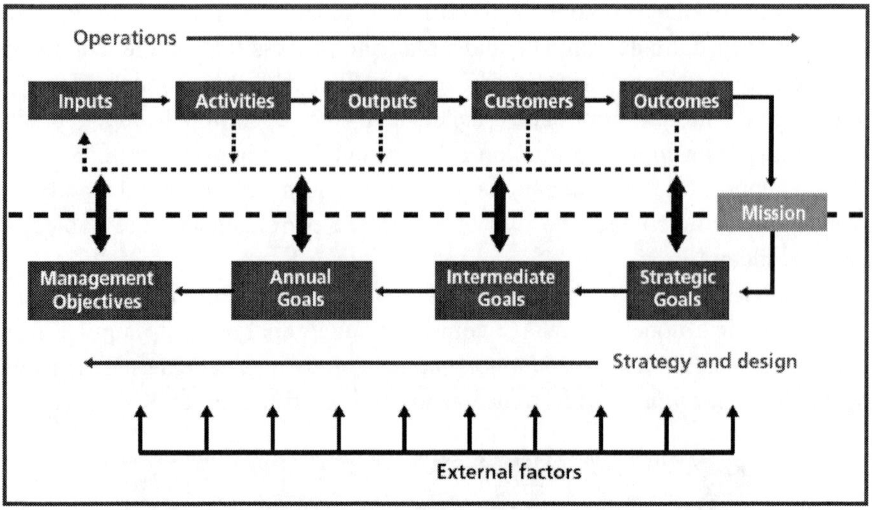

Figure 3. Schematic of the overall NIOSH logic model.

Intramural and extramural researchers present their project proposals, which receive appropriate internal and external review and are funded on the basis of merit.

The conduct of research *(activities)* produces "outputs" such as guidance documents and reports that address effective risk management practices, worker and employer education, and new technologies for assessing and controlling workplace hazards. NIOSH outputs are transferred directly to the final customers and stakeholders (who implement improvements in workplace safety and health) or to intermediate customers (who utilize NIOSH outputs to produce *intermediate outcomes*). These intermediate outcomes, such as training programs, regulations, and occupational standards, are used to advance workplace safety and health. Since NIOSH is not a regulatory agency, it relies heavily on efforts by intermediate and final customers to achieve ultimate outcomes in the form of workplace safety and health improvements. The effectiveness in achieving these outcomes is influenced at all stages of the program operation by both external factors (such as economic and social conditions) and the regulatory environment. Results of NIOSH-funded research and customer feedback (*intermediate and final outcomes*) contribute to the subsequent rounds of program planning.

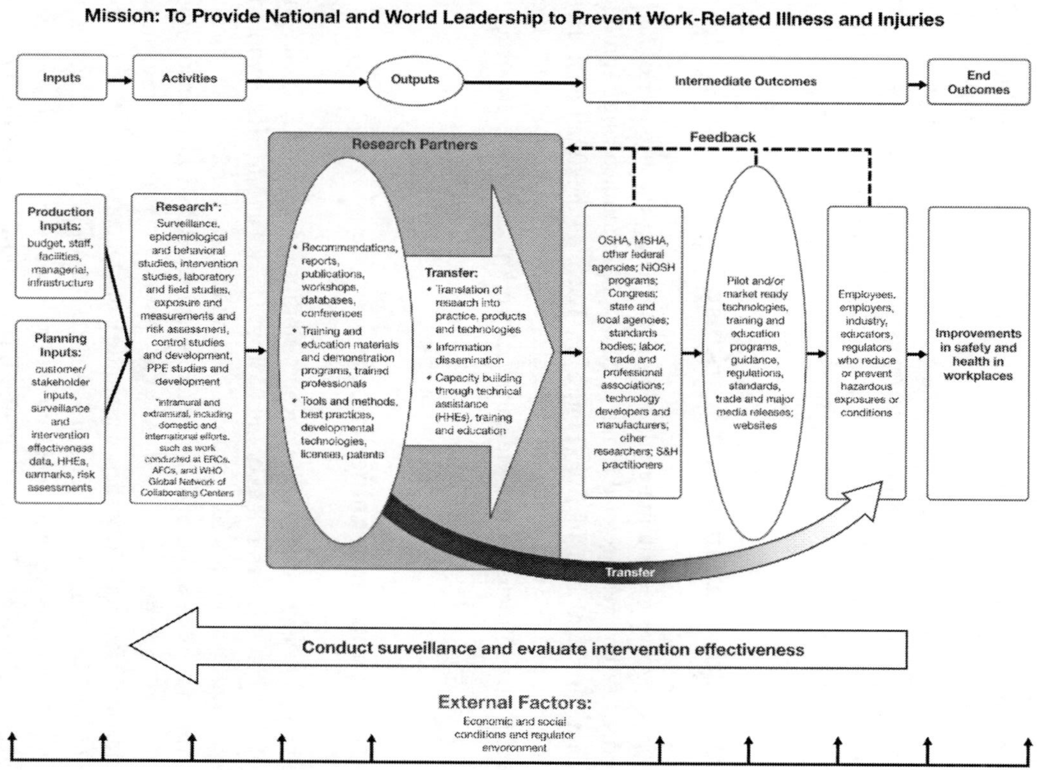

Figure 4. Schematic of the NIOSH operational model.

2. INPUTS

2.1. Nature of Nanomaterials

The universe of nanomaterials is potentially large because of the array of possible combinations of physicochemical parameters, impurities, and manufacturing and production conditions (Figure 5). NIOSH recognizes the extensive diversity and number of nanomaterials and the even larger number of applications for those materials. Consequently, it continues to use approaches that allow for a holistic view of nanomaterials and development of priorities that will be beneficial to many workers.

2.2. Nature of the Nanomaterial Workplace

Worker exposure to nanomaterials may occur along the life cycle of the nanomaterial, from research through scale-up, manufacturing, product development, use, and end of life (Figure 6). Each of the functional areas represents work-places with different potential exposure scenarios and risks. NIOSH continues to focus on the wide array of workplaces where exposure to nanomaterials may occur.

2.3. Customer and Stakeholder Input

To direct its research and fulfill its responsibilities under the OSH Act and the FMSH Act, NIOSH has relied on input from OSHA and MSHA, workers, employers, trade associations, unions, occupational safety and health practitioners and researchers, and the general public. NIOSH also seeks input through formal committees, such as the NIOSH Board of Scientific Counselors, the National Advisory Committee on Occupational Safety and Health, and the Mine Safety and Health Research Advisory Committee. It also receives input through ad hoc mechanisms such as the NIOSH Web site (www.cdc.gov/niosh), a toll-free telephone line (1-800-CDC-INFO), personal contacts with occupational safety and health professionals, and participation in professional conferences and interagency committees. In addition, NIOSH provides stewardship of the National Occupational Research Agenda (NORA) (http://www.cdc.gov/niosh/nora), which is a framework to guide occupational safety and health research into the new millennium—not only for NIOSH but for the entire occupational safety and health community.

NIOSH continues to be a participating and partnering agency in the National Nanotechnology Initiative (NNI) and provides occupational health and safety expertise in support of the U.S. Government's Strategic Plan for Nanotechnology Environmental, Health, and Safety Research [NNI 2011]. NIOSH is co-leader of the NNI Nanotechnology Environmental and Health Implications Working Group (NEHI) and contributed to the development of the 2011 NEHI Environmental, Health, and Safety (EHS) Research Strategy. The 2011 NEHI EHS research strategy identified five core nanotechnology EHS research categories: (1) nanomaterial measurement infrastructure, (2) human exposure assessment, (3) human health, (4) environment, and (5) risk assessment and management methods (Appendix A) [NNI 2011]. NIOSH provided significant input for the strategy and was the lead agency for developing recommendations on human exposure assessment. The NNI EHS research strategy focuses on the use of science-based risk analysis and risk management to protect public health and the environment while fostering the technological advancements that benefit society. NIOSH has strived to keep the overarching research activities of the NTRC nanotechnology strategic plan (Section 4.3) consistent with the NNI EHS research strategy.

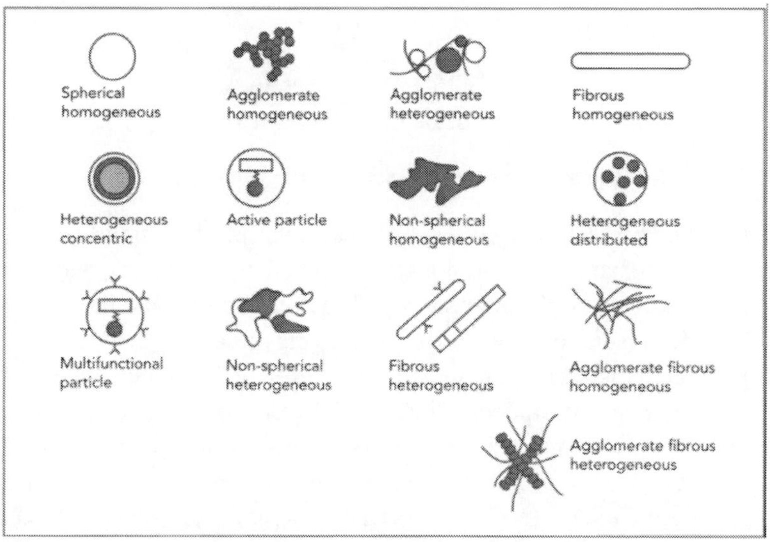

Figure 5. The wide variety of physicochemical parameters of engineered nanomaterials [Schulte et al. 2009].

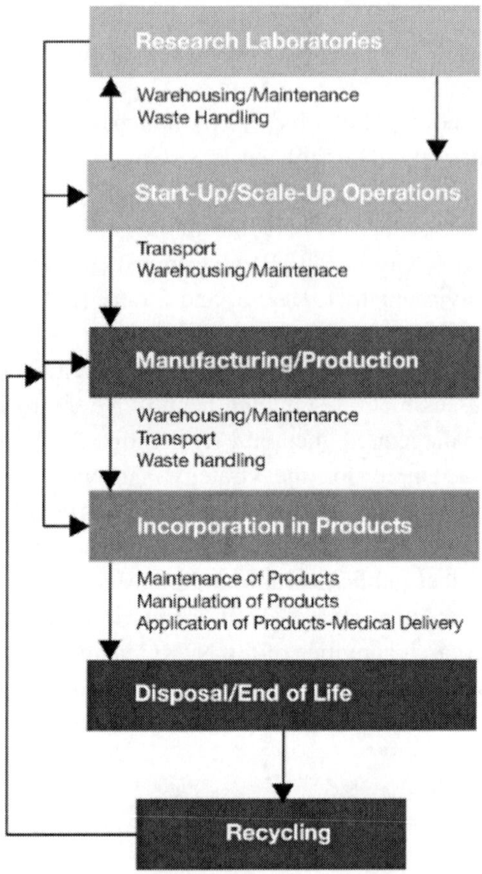

Figure 6. The life cycle of nanomaterials [Schulte et al. 2008a].

The NIOSH NTRC also fosters stakeholder in-put with trade associations, professional associations, labor, nongovernmental organizations, and a number of private nanomaterial companies. These collaborations have provided expertise and resources critical for reviewing research activities and for developing and disseminating health and safety information on engineered nanoparticles. Some of the ongoing NTRC stakeholders are the American Industrial Hygiene Association (AIHA), American Society of Safety Engineers (ASSE), International Safety Equipment Association (ISEA), National Safety Council (NSC), American Federation of Labor and Congress of Industrial Organizations (AFL–CIO), and NanoBusiness Commercialization Association (NanoBCA).

2.4. NIOSH Research Capabilities

Within its divisions, NIOSH has world-renowned researchers who are trained and experienced in the full spectrum of scientific skills needed to evaluate and characterize workplace hazards and devise intervention strategies for protecting workers. In addition to this expertise, NIOSH has significant laboratory capabilities in particle measurement, aerosol collection and characterization, particle surface analysis, measurement of particle surface radicals and activity, and in vitro and in vivo analysis of toxicity and pathogenesis. Laboratories conducting nanomaterial research are located in Cincinnati, Ohio; Pittsburgh, Pennsylvania; and Morgantown, West Virginia. NIOSH researchers work closely with a broad range of scientists from industry, academia, and other government agencies. The involvement of NIOSH in national and international initiatives and programs is an important component of its capacity to address critical occupational safety and health issues in nanotechnology.

2.5. NIOSH Partnerships

NIOSH recognizes both the practical need and the leadership obligation to extend its internal capabilities by leveraging activities and expertise found in other research institutions, industries, federal agencies, and nongovernmental organizations. These partnerships serve to deliver on multiple objectives; most important, they add to the body of knowledge on workplace health and safety issues associated with nanotechnology. Partnerships have taken several forms, ranging from formal letters or memoranda of understanding to informal working agreements on a specific topic. NIOSH will continue to pursue partnerships as a means of achieving the goals of this strategic research plan and as effective vehicles to develop and disseminate research results. The NIOSH NTRC has successfully used partnerships with industry to gain a better understanding of actual industrial nanomaterial processes, workplace exposures, work practices, and exposure control techniques. The field research conducted by the NTRC to assess exposures to engineered nanoparticles represents ongoing partnerships with numerous companies. The NTRC will continue to develop these partnerships to

- better understand how engineered nanoparticles are being produced and used
- develop recommendations for the safe handling of nanomaterials
- develop sampling and analytical methods
- evaluate exposure controls that are or could be used in nanomaterial processes
- evaluate the need for and determine the effectiveness of PPE, including respiratory protection
- develop communication and information materials that will assist industry in communicating with workers and the public.

Several of the industrial partnerships have provided opportunities for the NTRC to identify areas in which additional research is needed.

The continued collaboration with other national and international research institutes, academia, and government agencies provides the NTRC the opportunity to combine its expertise in workplace health and safety with the capabilities of other organizations to investigate specific areas of concern. Working relationships with other research institutes provide the NTRC with the information needed to guide its own research, focus its limited resources effectively, and expand its research capabilities through scientific collaboration. The NTRC has developed partnerships in the areas of toxicology, risk assessment modeling, exposure measurement methods, epidemiology, industrial hygiene, control technologies, filtration of nanoparticles, and communication of research results and safe work practices.

The NTRC has broadened its activities with collaborators and stakeholders by participating in a number of national and international committees and working groups. This participation provides NTRC scientists the opportunity to provide and receive input on the key research that is needed to address priority areas.

3. Activities

3.1. NIOSH Nanotechnology Research Center (NTRC)

Vision of the NTRC

The vision of the NTRC is as follows:

Safe nanotechnology by delivering on the Nation's promise—safety and health at work for all people through research and prevention.

Mission of the NTRC

The mission of the NTRC is to provide national and world leadership for research and guidance on the implications of nanoparticles and nanomaterials for work-related injury and illness, and the application of nanoparticles and nanomaterials in occupational safety and health.

3.2. NTRC Steering Committee

The NTRC Steering Committee is chaired by the NTRC Coordinator. The committee is made up of the program manager, program coordinator, program assistant coordinator, and critical area coordinators for each of the 10 research areas (some areas have multiple research coordinators). The steering committee is responsible for guiding NIOSH scientific and organizational plans in nanotechnology research (including coordination for science and budget) and for developing strategic goals and objectives and performance measures for the NTRC. Regular updates and progress reports on internal research activities are managed through biweekly teleconferences among members of the NTRC Steering Committee. In addition, to ensure the responsiveness, relevance, and impact of the NIOSH nanotechnology program, members of the NTRC and appropriate stakeholders meet in person every 2 years at a scientific exchange meeting.

3.3. Current NIOSH NTRC Intramural Nanotechnology Research Activities

Current NIOSH NTRC research activities in nanotechnology are focused on occupational safety and health implications and applications of ENMs. Data gathered from ongoing research studies with engineered nanoparticles are used to better understand workplace nanoparticle exposures, the hazards posed by nanomaterials, and the potential risk of adverse health effects from occupational exposures. Studies are also providing data on the characteristics of nanomaterials produced and used in the workplace, routes of exposure, and the effectiveness of work practices and engineering controls in preventing worker exposure. Findings from these intramural studies are providing scientific data to support the development of occupational safety and health recommendations. NIOSH progress and accomplishments in nanotechnology research can be found in the report *Filling the Knowledge Gaps for Safe Nanotechnology in the Work- place* [NIOSH 2012a].

3.3.1. Evaluation of the nanotechnology research program

NIOSH evaluates its research progress in nanotechnology through a series of internal and external reviews. Past reviews of its nanotechnology research program have been conducted by the National Academies and by a work-group of scientists chaired by two members of the NIOSH Board of Scientific Counselors. Based on these reviews, a revised strategic plan was published in November 2009 that proposed focused research and guidance on (1) identifying categories of nanomaterials that can be distinguished by their physical and chemical properties; (2) recommending specific exposure limits for various nanomaterials; (3) assessing hazards, risks, and exposures; and (4) developing intervention strategies [NIOSH 2009b]. In addition to external reviews, NIOSH continues to conduct annual program reviews and regular project reviews, and it publishes periodic progress reports describing ongoing research and accomplishments.

This *NIOSH Nanotechnology Research and Guidance Strategic Plan* was reviewed by a workgroup of scientists chaired by two members of the NIOSH Board of Scientific Counselors, and public comment was solicited through an announcement in the Federal Register.

3.4. Current NIOSH Extramural Nanotechnology Research Activities

The NIOSH Office of Extramural Programs (OEP) manages the competitive process for awarding occupational safety and health grants and cooperative agreements to the research community outside the Institute. This process involves peer review, program relevance, and priorities of the National Occupational Research Agenda (NORA), the NIOSH Research to Practice (r2p) initiative, congressional mandates, and sector, cross-sector, or coordinated emphasis areas of the NIOSH Program Portfolio (http://www. cdc.gov/niosh/programs).

Since 2001, the NIOSH OEP has funded nanotechnology research through Occupational Safety and Health Research Program Announcements (R01), Mentored Scientist Grants (K01), Small Research Grants (R03),

Developmental Grants (R21), and Small Business Innovation Research Grants (R43/44). During the period 2001 to 2012, the NIOSH OEP has committed approximately $12 million to extramural nanotechnology research.* Summaries of the projects funded by the NIOSH OEP are included in the document *Filling the Knowledge Gaps for Safe Nanotechnology in the Workplace* [NIOSH 2012a]. Through continued collaboration with the Environmental Protection Agency/National Center for Environmental Research (EPA/NCER), National Science Foundation (NSF), National Institutes of Health/National Institute of Environmental Health Sciences (NIH/NIEHS), and international agencies, the NIOSH OEP continues to support nanotechnology research that focuses on occupational safety and health issues. In addition, the NIOSH OEP routinely confers with the NIOSH NTRC regarding research needs.

4. Goals

4.1. Address Each Element in the Risk Management Continuum

Ultimately, the goal of the NTRC is to develop information, knowledge, and guidance to protect the nanotechnology workforce. It is relatively early in the emergence of nanotechnology and there are many gaps in knowledge, yet nano-enabled products are in commerce and workers are being exposed to nanomaterials. Consequently, there is a need to address all elements in the risk management continuum concurrently.

The process for managing potential workplace exposures to engineered nanoparticles during their synthesis, manufacture, and incorporation into new materials and devices consists of the following steps to ascertain the appropriate risk management strategy:

1. identifying and characterizing the health and safety hazard;
2. conducting dose-response risk assessment;
3. assessing the extent of exposure;
4. characterizing the risk on the basis of exposure; and
5. developing control and management procedures [Schulte et al. 2008b].

* This includes all extramural projects that involve any aspect of nanotechnology.

As exposure assessment data for nanomaterials become available, the existence of any occupational risk—and the conditions under which exposure to the hazard would be harmful to workers—can be determined. A goal of the risk characterization is to determine whether exposure to a type of material (in this case, nanomaterials) used in a given technology is likely to result in adverse health effects. Exposure assessment data provide a means to determine whether workers might be exposed and what type of control strategy might be effective in preventing exposure. Epidemiologic research is needed to provide etiologic evidence for new hazards and quantitative exposure–response information for recommended exposure limits. Decisions about which nanomaterials to study epidemiologically require collecting information on workforce size, exposure levels, and expected hazard (from toxicological or other information). The NIOSH NTRC is involved in conducting research and answering questions posed in each element of the risk management process to address knowledge gaps in the protection of workers (Figure 7).

A challenge of the NIOSH NTRC program is to determine how to conduct timely research that addresses the elements of hazard identification through risk management, as introduction of nanomaterials to the workplace continues and the workforce exposed to them becomes more diversified. The approach taken by the NTRC has been to conduct concurrent, focused research to address knowledge gaps in each step of the risk management process to provide occupational safety and health guidance.

4.2. Identification of NTRC Data Needs for Knowledge Gaps

In preparing this strategic plan, the NTRC steering committee met in spring 2012 to identify data needs and research gaps. Information was gleaned from various sources:

- research studies conducted by the NIOSH NTRC and others
- comments received during public review from open meetings or submitted to the NIOSH Docket on *Interim Guidance for Medical Screening and Hazard Surveillance for Workers Potentially Exposed to Engineered Nanoparticles*
- comments received during public review from open meetings or submitted to the NIOSH Docket on *Current Intelligence Bulletin for Occupational Exposure to Carbon Nanotubes and Nanofibers*

- comments received in the NIOSH Docket on the draft strategic goals [76 Fed. Reg.* 123612]
- the 2011 NEHI Environmental, Health, and Safety (EHS) Research Strategy
- various scientific meetings and collaborations, including tripartite attendance (labor, industry, and government) and stakeholder input workshops.

On the basis of this information, the following data needs and research gaps have been identified.

Increase understanding of new hazards and related health risks to nanomaterial workers

- Identify emerging commercial ENMs through market forecasting and research, technology surveillance, and partner and stakeholder input.
- Determine how workers might be exposed to nanoparticles during manufacturing, handling, and use of nanomaterials.
    - Expand research from primary manufacturer to second-, third-, and fourth-generation use of ENMs.
    - Continue field investigations to better understand worker exposures throughout the life cycle, including processing of nanoenabled products.
    - Determine how ENMs interact with the body's systems.
    - Determine relationships between specific physicochemical properties of ENMs and adverse biological effects.

Expand understanding of the initial hazard findings of engineered nanomaterials

- Determine carcinogenicity of carbon nanotubes (CNTs).
- Determine whether the cardiovascular system is more sensitive than the lungs to effects from ENM exposures.
- Determine whether ENMs cause immune dysfunction.
- Explore the usefulness of biomarkers for identifying and determining the extent of worker exposure and the potential for adverse health effects.

* Federal Register.

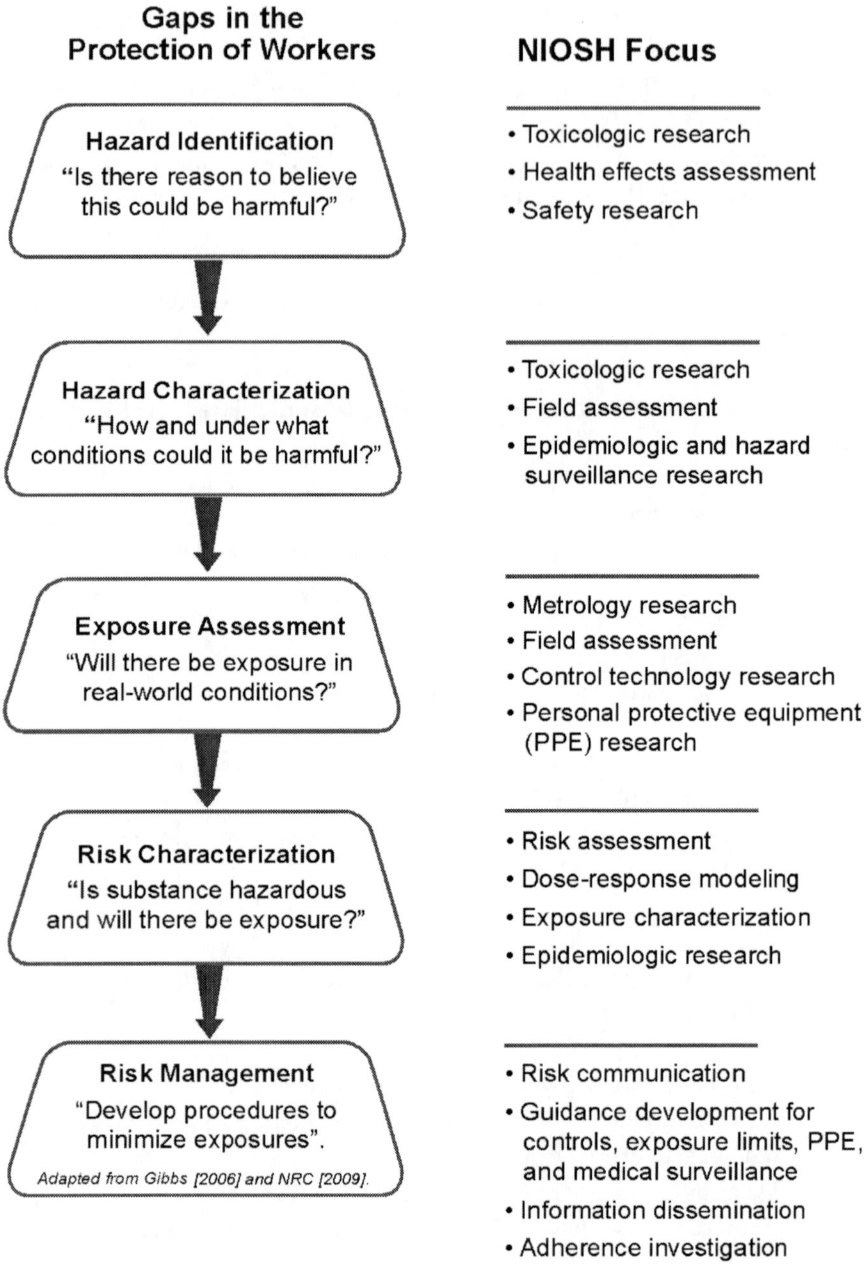

Figure 7. Addressing knowledge gaps in protecting nanomaterial workers.

- Conduct research on measurement methods for nanomaterials.
  - Align laboratory and field exposure metrics and measurement techniques.
  - Expand and refine workplace measurement techniques for assessing exposure (tiered approach).
  - Develop a microscopy method for CNTs and other ENMs.
  - Identify the critical reference materials needed for measurement.
- Continue to evaluate additional ENMs for dustiness, explosibility, flammability, and electromagnetic hazard potential.
  - Correlate dustiness, explosion, fire, and electromagnetic data with existing hazard indexes.

Support the creation of guidance materials to inform nanomaterial workers, employers, health professionals, regulatory agencies, and decision-makers about the hazards, risk, and risk management approaches.

- Develop categorical and specific occupational exposure limits.
  - Conduct research to recommend specific exposure limits for various ENMs (e.g., nanosilver, graphene, carbon black, nano silica, nanoceramics, nano clays, nano catalysts), and evaluate the adequacy of existing exposure limits.
- Identify categories of nanomaterials that can be distinguished by physicochemical parameters.
- Conduct risk assessments.
  - Nanoparticle lung dosimetry modeling
  - Categorization and comparative potency methods
- Link risk assessment, toxicology, exposure measurement, and control strategies.
- Demonstrate effective risk management practices.
- Develop guidance materials.
  - Update guidance for medical surveillance (include sensitivity and specificity of medical tests).
  - Evaluate and provide guidance on control technology.
  - Integrate Prevention through Design (PtD) into ENM risk management guidelines.
  - Use categorical approaches for ENMs to develop guidance on controlling exposures (hazard and control banding).

    - Evaluate and provide guidance on PPE for use with ENMs.
- Partner with businesses to promote safe and responsible handling of ENMs.
- Determine whether there are potential applications of nanomaterials for safety and health (e.g., improved sensors or PPE).

Support epidemiologic studies for nanomaterial workers, including medical, cross-sectional, prospective cohort and exposure studies.

- Conduct epidemiologic research on the health of nanomaterial workers by employing both cross-sectional and prospective designs.
- Partner with other U.S. agencies and groups that have begun epidemiologic and other surveillance studies.
- Determine appropriate tools for medical screening and mechanisms of medical surveillance among workers exposed to specific EMNs.
- Create an exposure database that harmonizes with global exposure databases.

Assess and promote national and international adherence with risk management guidance

- Conduct a national assessment to determine the extent of use of nanomaterial exposure controls and any barriers to their use.
- Provide intervention strategies for industry sectors where adherence to good risk- management practices is low.
- Partner with other businesses, U.S. agencies, and the international community to promote safe, responsible development of nanomaterials.

4.3. Coalescing Priorities for NIOSH NTRC Research and Guidance for FY2013–FY2016

The NIOSH NTRC has proposed specific research activities (see Section 4.5) for FY2013– FY2016 that address the previously identified research needs and gaps (Section 4.2). Figure 8 provides a schema for how NIOSH will focus its research activities in accordance with the goals of the NTRC Strategic Plan:

- Increase understanding of new hazards and related health risks to nanomaterial workers.
- Expand understanding of the initial hazard findings of engineered nanomaterials.
- Support the creation of standards and guidance materials to inform nanomaterial workers, employers, health professionals, regulatory agencies, and decision-makers about hazards, risks, and risk management approaches.
- Support epidemiologic studies for nanomaterial workers, including medical, cross-sectional, prospective cohort, and exposure studies.
- Assess and promote national and international adherence with risk management guidance.

The overall focus of the strategic plan will be to continue to generate data, information, and knowledge to protect the nanotechnology workforce, by using effective risk management techniques and particularly keeping exposure below occupational limits. Research is proposed to develop and update good risk-management practices, including categorical and specific exposure limits. Research data and information will be cataloged and shared via informatics (resources, devices, and methods required to optimize the acquisition, storage, retrieval, and use of information).

4.4. Coordination with the National Nanotechnology Initiative

NIOSH research priorities address many of the strategic goals proposed by the NNI Strategy for Nanotechnology Environmental, Health, and Safety Research. Table 1 shows the alignment of the NIOSH NTRC strategic goals and 10 NTRC critical research areas with the NNI EHS priority research needs (Appendix A).

4.5. Proposed NIOSH NTRC Research

The following proposed research for FY2013– FY2016 has been coalesced to focus on specific research needs to fill data gaps (see Section 4.2) and the EHS priority research needs of the NNI (see Section 4.4). As knowledge gaps are filled and as resources become available, these research priorities could change.

Strategic Goal 1 (PPNANSG1): Increase understanding of new hazards and related health risks to nanomaterial workers.

Intermediate Goal 1.1 (PPNANIG1.1): Conduct market forecasting and review available surveillance data to determine emerging commercial ENMs and next-generation uses of existing ENMs.

Activity/Output Goal 1.1.1 (PPNANAOG1.1.1): Prioritize high-volume emerging ENMs to identify the next candidates for toxicological testing and field evaluation of workplace exposures.

Performance Measure 1.1.1: In FY2013, complete an initial market research and forecasting report for internal use. Update the report within 3 years.

Activity/Output Goal 1.1.2 (PPNANAOG1.1.2): Complete a secondary market analysis on high volume nanomaterials.

Performance Measure 1.1.2: By FY2014, complete a market landscape analysis of formulators and users of nano titanium dioxide, silver, graphene and cellulose. Identify key industries using these materials in current products and those to be commercialized in the near term.

Intermediate Goal 1.2 (PPNANIG1.2): Conduct research to contribute to the understanding of the toxicology and internal dose of emerging ENMs.

Activity/Output Goal 1.2.1 (PPNANAOG1.2.1): Evaluate acute and chronic effects in the lungs and in other organ systems and tissues. Determine dose-response and time-course relationships. Determine rates of clearance of these nanoparticles after pulmonary exposure and translocation to systemic organs; characterize systemic effects.

Performance Measure 1.2.1: Complete toxicologic studies on ENMs such as nanosilver, graphene, and nanocellulose; present the data via the Society of Toxicology and publish in the peer-reviewed literature by FY2016.

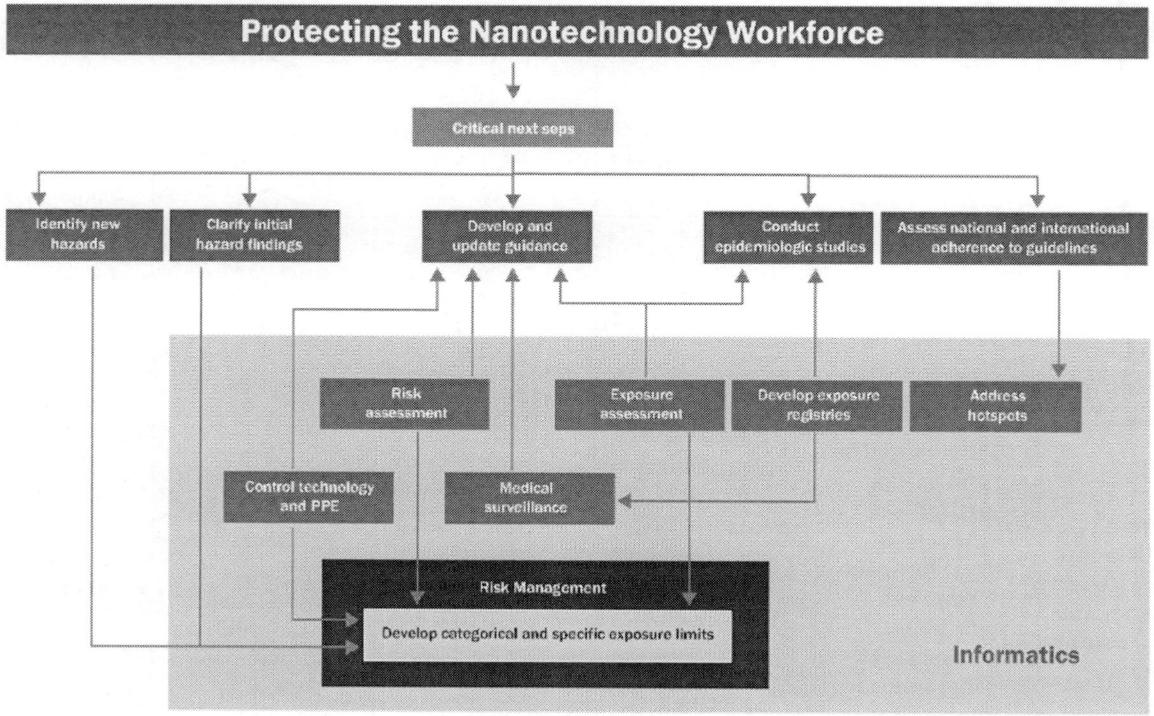

Figure 8. Focus of NIOSH nanomaterial research, 2013–2016.

Table 1. Alignment of Critical Research Areas with the Five Strategic Goals of the NIOSH NTRC and the NNI EHS Priority Research Needs (RNs)

NIOSH NTRC Strategic Goal/ Critical Research Area	Increase understanding of new hazards and related health risks to nanomaterial workers	Expand understanding of the initial hazard findings of engineered nanomaterials	Support the creation of guidance materials to inform nanomaterial workers and employers about hazards, risks, and risk management approaches	Support epidemiologic studies for nanomaterial workers, including medical, cross–sectional, prospective cohort, and exposure studies	Assess and promote national and international adherence with risk management guidance
Toxicity and internal dose	✓* 3-RN1† 3-RN2 3-RN3 3-RN4 3-RN5 4-RN1	✓ 3-RN1† 3-RN2 3-RN3 3-RN4 3-RN5	✓		
Measurement methods	✓ 1-RN1 1-RN2 1-RN3 2-RN1 3-RN2	✓ 1-RN1 1-RN2 1-RN3 2-RN1 3-RN2	✓		

NIOSH NTRC Strategic Goal/ Critical Research Area	Increase understanding of new hazards and related health risks to nanomaterial workers	Expand understanding of the initial hazard findings of engineered nanomaterials	Support the creation of guidance materials to inform nanomaterial workers and employers about hazards, risks, and risk management approaches	Support epidemiologic studies for nanomaterial workers, including medical, cross–sectional, prospective cohort, and exposure studies	Assess and promote national and international adherence with risk management guidance
Exposure assessment	✓ 2-RN1 2-RN3 3-RN2 4-RN1 5-RN2	✓ 2-RN1 2-RN3 3-RN2 4-RN1 5-RN2	✓		
Epidemiology and surveillance	✓	✓	✓	✓ 2-RN1 2-RN3 2-RN4 3-RN6 5-RN2	✓ 5-RN2
Risk assessment	✓ 5-RN1 5-RN2 5-RN5	✓ 5-RN1 5-RN2 5-RN5	✓ 5-RN3 5-RN4		

Table 1. (Continued)

NIOSH NTRC Strategic Goal/ Critical Research Area	Increase understanding of new hazards and related health risks to nanomaterial workers	Expand understanding of the initial hazard findings of engineered nanomaterials	Support the creation of guidance materials to inform nanomaterial workers and employers about hazards, risks, and risk management approaches	Support epidemiologic studies for nanomaterial workers, including medical, cross–sectional, prospective cohort, and exposure studies	Assess and promote national and international adherence with risk management guidance
Engineering controls and PPE	✓ 2-RN1 5-RN1 5-RN2	✓ 2-RN1 5-RN1 5-RN2	✓		✓ 5-RN2
Fire and explosion safety	✓ 1-RN5	✓ 1-RN5	✓		
Recommendations and guidance			✓ 5-RN‡ 5-RN§ 5-RN¶		✓ 5-RN2
Global collaborations‡	✓	✓	✓	✓	✓§
Applications¶			✓		

* A check mark (✓) indicates that a goal is currently being addressed by projects within the NIOSH critical research area.

† Alphanumeric identifications indicate specific NNI priority research needs (see Appendix A) being met by the NIOSH projects in each critical research area.

‡ International engagement is a priority of the NNI and a critical component of the 2011 NNI EHS Research Strategy. However, it is not identified as a research need in the NNI Strategy.

§ NIOSH will work with other governments to determine global adherence.

¶ The NNI EHS plan does not address applications of nanomaterials for EHS use. However, the NNI strategic plan includes applications as a driving force behind several Federal nanotechnology programs, and it has the potential to improve safety and health, such as through the development of advanced sensors or PPE.

Performance Measure 1.2.2: Identify probable combinations of nanomaterial uses and if frequent combinations are identified, conduct preliminary toxicologic evaluations to determine if synergistic effects occur by FY2016.

Intermediate Goal 1.3 (PPNANIG1.3): Determine whether nanomaterial toxicity can be categorized on the basis of physicochemical properties and mode of action.

Activity/Output Goal 1.3.1 (PPNANAOG1.3.1): Systematically investigate the physical and chemical properties of particles that influence their toxicity (e.g., size, shape, surface area, solubility, chemical properties, and trace components).

Activity/Output Goal 1.3.2 (PPNANAOG1.3.2.): Determine the biological mechanisms for toxic effects (e.g., oxidant stress, dissolution, fibrogenicity, and hydrophobicity) and how the key chemical and physical factors may influence these mechanisms.

Activity/Output Goal 1.3.3 (PPNANAOG1.3.3): Integrate mechanistic models (including animal models and in vitro screening tests) for assessing the potential toxicity of new nanomaterials, and provide a basis for developing predictive algorithms for structure/function relationships and comparative toxicity analyses for risk assessment. Determine whether toxicologic data from new studies can be linked to existing toxicologic data.

Performance Measure 1.3.1: By FY2016, determine whether ENMs can be categorized (hazard banded) on the basis of physicochemical properties and/or the toxicologic mode of action.

Intermediate Goal 1.4 (PPNANIG1.4): Conduct nanomaterial workplace exposure assessments at sites developing, producing, or using emerging ENMs.

Activity/Output Goal 1.4.1 (PPNANAOG1.4.1): Collaborate with nanomaterial businesses to identify processes that may release nanomaterials during manufacturing or use of ENMs.

Evaluate potential exposures and document risk mitigation processes.

Performance Measure 1.4.1.1: Complete a minimum of four field evaluations per year at facilities producing or using emerging ENMs. By FY2014, publish summary findings in a peer-reviewed journal.

Performance Measure 1.4.1.2: Evaluate ENM use and risk management practices at user facilities across the life cycle, including secondary and tertiary facilities. By FY2015, publish summary findings in a peer-reviewed journal.

Strategic Goal 2 (PPNANSG2): Expand understanding of the initial hazard findings of engineered nanomaterials.

Intermediate Goal 2.1 (PPNANIG2.1): Conduct research to evaluate the adverse chronic effects of single-walled carbon nanotubes, multi-walled carbon nanotubes, and carbon nanofibers.

Activity/Output Goal 2.1.1 (PPNANAOG2.1.1): Conduct toxicologic testing to determine the potential of single-walled carbon nanotubes, multi-walled carbon nanotubes, and carbon nanofibers to have long-term pulmonary effects (fibrosis, lung cancer, mesothelioma).

Performance Measure 2.1.1: By FY2016, publish a report on the carcinogenicity of carbonaceous nanomaterials, based on completed carcinogenicity-toxicology studies.

Intermediate Goal 2.2 (PPNANIG2.2): Determine whether human biomarkers of nanomaterial exposure and/or response can be identified.

Activity/Output Goal 2.2.1 (PPNANAOG2.2.1): Conduct research to determine whether blood, urine, or nasal lavage markers can accurately predict initiation and progression of adverse responses to ENMs.

Performance Measure 2.2.1: By FY2017, publish results of human biomarker studies.

Activity/Output Goal 2.2.2 (PPNANAOG2.2.2): Conduct research to determine whether cardiovascular or immune responses are sensitive indicators of pulmonary exposure to ENMs.

Performance Measure 2.2.2: By FY2016, publish findings on cardiovascular and immune responses to pulmonary exposure to ENMs.

Intermediate Goal 2.3 (PPNANIG2.3): Determine the relevance of in vitro and in vivo screening tests to worker response to inhalation of ENMs.

Activity/Output Goal 2.3.1 (PPNANAOG 2.3.1): Evaluate the relationship of deposition patterns and biological responses to pulmonary bolus exposure (intratracheal instillation or pharyngeal aspiration) vs. exposure extended over time (inhalation) of ENMs in rodent models.

Performance Measure 2.3.1: By FY2016, publish findings on the relationship of deposition patterns and biological responses to pulmonary bolus exposure vs. exposure extended over time.

Activity/Output Goal 2.3.2 (PPNANAOG 2.3.2): Evaluate predictive in vitro screening tests for fibrogenicity, genotoxicity, and/or cell transformation.

Performance Measure 2.3.2: Publish results from predictive in vitro screening tests for fibrogenicity, genotoxicity, and/or cell transformation, over the next 2 to 4 years (FY2014–FY2016).

Intermediate Goal 2.4 (PPNANIG2.4): Conduct research on measurement methods for nanomaterials.

Activity/Output Goal 2.4.1 (PPNANAOG2.4.1): Continue interactions with national metrology institutes and qualify cellulose and single-walled CNT nanoscale reference materials (RMs) and/ or benchmark materials for evaluating measurement tools, instruments, and methods for exposure assessment and toxicology.

Performance Measure 2.4.1.1: By FY2014, complete a round-robin evaluation of a transmission electron microscopy (TEM) sample-preparation procedure using National Institute of Standards and Technology (NIST) gold nanoparticle RMs 8011, 8012, and 8013.

Performance Measure 2.4.1.2: By FY2014, translate this TEM sample-preparation procedure into a consensus standard practice or guide through ASTM International Committee E56: Nanotechnology.

Performance Measure 2.4.1.3: Within 4 years (FY2016), utilize NIST gold nanoparticle RMs 8011, 8012, and 8013 to evaluate filter- and grid-based sample-collection techniques and TEM analysis for spherical ENMs.

Performance Measure 2.4.1.4: By FY2016, develop measurement protocols for dynamic light scattering (hydrodynamic diameter), laser Doppler electrophoresis (zeta potential), and gas adsorption (surface area) measurements that utilize RMs (as appropriate) for particle characterization for internal use to promote reproducibility of testing.

Performance Measure 2.4.1.5: By FY2016, develop instrument sampling cards that provide a succinct summary on the operation and utility of condensation particle counters, scanning mobility particle sizers, and diffusion chargers for internal use to promote measurement quality.

Activity/Output Goal 2.4.2 (PPNANAOG2.4.2): Understand relationships between laboratory instrumentation used to characterize aerosols for toxicology studies and field-measurement instrumentation used to characterize workplace atmospheres.

Performance Measure 2.4.2: By FY2016, complete a series of studies to evaluate test atmospheres with real-time and integrated laboratory and field-measurement instruments for a range of metrics, including particle number, size, and surface area.

Activity/Output Goal 2.4.3 (PPNANAOG2.4.3): Conduct further research on the dustiness testing of powders and to benchmark the Venturi (University of North Carolina) dustiness testing device against a more traditional dustiness test (e.g., a standard method prescribed in EN 15051).

Performance Measure 2.4.3: By FY2015, compare the Venturi dustiness testing device against at least one other method for a range of fine and nanoscale materials.

Activity/Output Goal 2.4.4 (PPNANAOG2.4.4): Conduct further research on determining the advantages and limitations of direct reading instruments in assessing nanoparticle work-place emissions and exposures.

Performance Measure 2.4.4: By FY2015, publish at least one further study on the performance of direct reading instruments for assessing nanoparticle emissions or exposures in the workplace.

Activity/Output Goal 2.4.5 (PPNANAOG2.4.5): Develop and improve methods and approaches (including direct reading and time-integrated sampling) for assessing workplace exposures to ENMs.

Performance Measure 2.4.5.1: By FY2015, publish updated strategies for assessing work-place exposures.

Performance Measure 2.4.5.2: By FY2015, validate method 5040 for a series of CNT including. NIST Standard Reference Material 248: Single-Wall Carbon Nanotubes (Raw Soot).

Activity/Output Goal 2.4.6 (PPNANAOG2.4.6): Evaluate scanning electron microscopy, transmission electron microscopy, and dark-field microscopy as analytical tools for measurement of carbon nanotubes using NIST Standard Reference Material 248: Single-Wall Carbon Nanotubes (Raw Soot).

Performance Measure 2.4.6: By FY2015, publish a microscopy method (or methods) for sample preparation, identification, and/or possible semi-quantification that is applicable to some ENMs.

Activity/Output Goal 2.4.7 (PPNANAOG2.4.7): Conduct research to advance the use of nanomaterials in the development of sensors (e.g., detection of harmful chemicals) and other health and safety–related applications that can be used in the workplace to ensure effective control of exposure.

Performance Measure 2.4.7.1: By FY2016, develop an effective sensor employing nanomaterials for the real-time measurement of a workplace agent, and publish the results.

Performance Measure 2.4.7.2: By FY2016, compile and publish a compendium on applications of nanotechnology to improve work-place safety and health.

Intermediate Goal 2.5 (PPNANIG2.5): Conduct research to better define the potential fire and explosion safety hazards of ENMs.

Activity/Output Goal 2.5.1 (PPNANAOG2.5.1): Identify physical and chemical properties that contribute to dustiness, combustibility, flammability, explosibility, and electromagnetic hazards of ENMs.

Performance Measure 2.5.1: By FY2014, publish preliminary findings from explosibility testing of some ENMs.

Performance Measure 2.5.2: By FY2016, publish findings that may influence the National Fire Protection Association (NFPA) or Department of Transportation (DOT) flammability placard information and the DOT Emergency Response Guidebook.

Intermediate Goal 2.6 (PPNANIG2.6): Develop and use informatics to process and communicate information.

Activity/Output Goal 2.6.1 (PPNANAOG2.6.1): Create a harmonized database to share exposure measurement, control, and epidemiologic data.

Performance Measure 2.6.1: By FY2015, have a populated dataset containing NIOSH results on measurement, control, and epidemiologic data.

Activity/Output Goal 2.6.2 (PPNANAOG2.6.2): Lead the nanoinformatics community to develop an open literature monograph on nanoinformatics principles and practice.

Performance Measure 2.6.2: Complete and publish the nanoinformatics monograph by FY2015.

Strategic Goal 3 (PPNANSG3): Support the creation of guidance materials to inform nanomaterial workers, employers, health professionals, regulatory agencies, and decision-makers about hazards, risks, and risk management approaches.

Intermediate Goal 3.1 (PPNANIG3.1): Develop categorical and specific exposure limits.

Activity/Output Goal 3.1.1(PPNANAOG3.1.1): Conduct a risk assessment for high-volume ENMs.

Performance Measure 3.1.1: By FY2015, complete a quantitative risk assessment (QRA) on ultrafine and fine materials from existing studies. Evaluate QRA methods for nanomaterials. Start QRA for nanoparticles, with use of new NIOSH data. Use NIOSH nanoparticle data to calibrate and validate dosimetry models for nanoparticles.

Activity/Output Goal 3.1.2 (PPNANAOG3.1.2): Develop a risk-assessment framework for evaluating the hazard and predicting the risk of exposure to ENMs.

Performance Measure 3.1.2: By 2017, develop a risk-assessment framework to rank hazard and estimate risk from exposure to selected ENMs in the workplace.

Activity/Output Goal 3.1.3 (PPNANAOG3.1.3): Use a nanomaterial hazard banding classification scheme to group ENMs.

Performance Measure 3.1.3: By FY2016, develop a hazard banding classification scheme for ENMs on the basis of toxicological, chemical, and physical properties.

Intermediate Goal 3.2 (PPNANIG3.2): Conduct research to better understand engineering controls and PPE for use with ENMs.

Activity/Output Goal 3.2.1 (PPNANAOG3.2.1): Evaluate the effectiveness of engineering control techniques for ENMs and develop new approaches as needed.

Performance Measure 3.2.1: Conduct field investigations of workplaces where ENMs are manufactured and used to evaluate existing engineering controls and make recommendations on improving exposure control. By FY2015, publish case studies and workplace survey reports, and by FY 2016, publish updated engineering control solutions.

Activity/Output Goal 3.2.2 (PPNANAOG3.2.2): Conduct laboratory evaluations of commercially available engineering controls for a variety of common ENM production, handling, and downstream processes.

Performance Measure 3.2.2: By FY2016, publish in-depth reports and journal articles on the effectiveness of the commercially available engineering control devices.

Activity/Output Goal 3.2.3 (PPNANAOG3.2.3): Evaluate the effectiveness of PPE (respirators, and protective clothing including gloves) for reducing worker exposures to ENMs.

Performance Measure 3.2.3: By FY2016, publish updated guidance on the effectiveness of PPE for reducing worker exposures to ENMs.

Intermediate Goal 3.3 (PPNANIG3.3): Incorporate Prevention through Design (PtD) into nanomaterial health and safety programs.

Activity/Output Goal 3.3.1 (PPNANAOG3.3.1): Promote PtD principles for nanomaterials, including safer nanomaterials that have the same functionality; process containment and control; and management system approaches to include occupational safety and health into the nanoparticle synthetic process, product development, and product manufacture.

Performance Measure 3.3.1: By FY2015, publish a document specific to PtD for nanomaterials.

Intermediate Goal 3.4 (PPNANIG3.4): Foster the collection, management, and dissemination of relevant information to protect nanomaterial workers.

Activity/Output Goal 3.4.1 (PPNANAOG3.4.1): Develop and disseminate effective information, education, and training materials to various target audiences such as nanotechnology workers and employers, occupational safety and health professionals, policy-makers, decision-makers, and/or the scientific community.

Performance Measure 3.4.1.1: By FY2015, create nanomaterial guidance documents for workers, including small business and construction workers.

Performance Measures 3.4.1.2: By FY2015, update *Approaches to Safe Nanotechnology: Managing the Safety and Health Concerns Associated with Engineered Nanomaterials*.

Performance Measures 3.4.1.3: By FY2016, update the *Progress Report on the NIOSH Nanotechnology Research and Communication Effort.*

Performance Measure 3.4.1.4: Create new guidance documents on use of engineering controls, PPE, and PtD, as per Performance Measures 3.2.1, 3.2.2, 3.2.3, and 3.3.1.

Performance Measure 3.4.1.5: By FY2016, support a conference on risk management of ENMs, with a focus on engineering controls.

Performance Measure 3.4.1.6: By FY2016, evaluate Nanomaterial Safety Data Sheets for adherence with the updated OSHA HazCom rule (including the use of Global Harmonization System), and issue guidance documents based on the updated findings.

Intermediate Goal 3.5 (PPNANIG3.5): Enhance global workplace safety and health through international activities.

Activity/Output Goal 3.5.1 (PPNANAOG3.5.1): Establish and maintain national and international partnerships so that knowledge gaps, research needs and priorities, approaches, and databases could be shared.

Performance Measure 3.5.1.1: Strengthen co-ordination of research through government-level organizations such as the Organization for Economic Cooperation and Development (OECD) and United Nations (UN).

Performance Measure 3.5.1.2: By FY2015, expand collaborations to developing nations.

Activity/Output Goal 3.5.2 (PPNANAOG3.5.2): Improve sharing critical data globally.

Performance Measure 3.5.2.1: By FY016, develop a global portal for information on ENMs relevant to occupational safety and health.

Performance Measure 3.5.2.2: If feasible, by FY2016 initiate the development of a global exposure registry database.

Performance Measure 3.5.2.3: In FY2015, participate in OECD Nanomaterial Safety Testing Program by sponsoring nanomaterial testing and by data exchange.

Activity/Output Goal 3.5.3 (PPNANAOG 3.5.3): Lead the development of global standards on occupational safety and health for nanotechnology.

Performance Measure 3.5.3.1: Over the next 3 years (FY2014–FY2016), lead the development of at least one standard in the International Organization for Standardization.

Performance Measure 3.5.3.2: Over the next 3 years (FY2014–FY2016), lead the development of at least one guidance document for the UN.

Performance Measure 3.5.3.3: Over the next 3 years (FY2014–FY2016), lead the development of at least one guidance document for the OECD.

Strategic Goal 4 (PPNANSG4): Support epidemiologic studies for nanomaterial workers, including medical, cross-sectional, prospective cohort, and exposure studies.

Intermediate Goal 4.1 (PPNANIG4.1): Conduct epidemiologic research and evaluate feasibility of conducting surveillance of nanomaterial workers.

Activity/Output Goal 4.1.1 (PPNANAOG4.1.1): Complete the epidemiological health studies of U.S. workers exposed to CNTs and carbon nanofibers.

Performance Measure 4.1.1: By FY2016, complete data collection for industry-wide exposure and epidemiological studies of workers exposed to carbonaceous nanomaterials.

Activity/Output Goal 4.1.2 (PPNANAOG4.1.2): Evaluate the need for and feasibility of initiating worker exposure registries for workers exposed to existing nanomaterials (e.g., titanium dioxide, CNTs) or producing and using new nanomaterials.

Performance Measure 4.1.2: By FY2016, develop a template and begin population of worker exposure registries.

Activity/Output Goal 4.1.3 (PPNANAOG4.1.3): Integrate nanotechnology safety and health guidance into existing hazard surveillance systems. Determine whether these systems are adequate by conducting evaluations.

Performance Measure 4.1.3: By FY2016, update *NIOSH Current Intelligence Bulletin 60: Interim Guidance for the Medical Screening and Hazard Sur veillance of Workers Potentially Exposed to Engineered Nanoparticles* [NIOSH 2009d].

Strategic Goal 5 (PPNANSG5): Assess and promote national and international adherence with risk management guidance.

Intermediate Goal 5.1 (PPNANIG5.1): Determine the extent to which control measures are being adopted by industry. Identify any barriers to use of controls and partner with business to address barriers.

Activity/Output Goal 5.1.1 (PPNANAOG 5.1.1): Determine whether NIOSH or other precautionary guidance recommendations are being adopted by businesses.

Performance Measure 5.1.1: Within the next 3 years, conduct formal and informal assessments of the controls used in ENM-producing or -handling facilities. By FY2016, publish findings.

Activity/Output Goal 5.1.2 (PPNANAOG 5.1.2): Develop a plan for industrial sectors where adherence to good risk-management practice is low.

Performance Measure 5.1.2: By 2016, publish a plan for intervening in industry sectors with low adherence to risk management guidance.

Activity/Output Goal 5.1.3 (PPNANAOG 5.1.3): Develop a follow-back program for the field investigation activities.

Performance Measure 5.1.3: By 2014, have a follow-back program in place to determine if facilities are implementing suggestions from the field investigations.

Activity/Output Goal 5.1.4 (PPNANAOG 5.1.4): Support the International Commission on Occupational Health (ICOH) and other organizations in assessing whether precautionary guidance recommendations are being adopted worldwide.

Performance Measure 5.1.4: By 2016, support ICOH in developing a plan to address locations where adherence to good risk-management practice is low.

5. Outputs

A summary of NIOSH NTRC research out-puts and accomplishments is presented in the report *Filling the Knowledge Gaps for Safe Nanotechnology in the Workplace* [NIOSH 2012a].

5.1. NIOSH Publications on Nanotechnology Since 2009

NIOSH NTRC scientists will continue to develop documentation that provides guidance and technical information for workers, employers, health professionals, regulatory agencies, and decision-makers in government, academia, industry, and labor. Examples of such documentation include the following:

- *Current Intelligence Bulletin (CIB) 65, Occupational Exposure to Carbon Nanotubes and Carbon Nanofibers*, DHHS (NIOSH) Publication No. 2013–145 [NIOSH 2013]. This document includes a recommended exposure limit for carbon nanotubes and nanofibers. It is available at http://www.cdc. gov/niosh/docs/2013-145/.
- *Filling the Knowledge Gaps for Safe Nanotechnolog y in the Workplace*, DHHS (NIOSH) Publication No. 2013–101 [NIOSH 2012a]. It is available at http:// www.cdc.gov/niosh/docs/2013-101/.
- *Current Intelligence Bulletin (CIB) 63, Occupational Exposure to Titanium Dioxide*, DHHS (NIOSH) Publication No. 2011–160 [NIOSH 2011]. This document includes a recommended exposure limit for both fine and ultrafine (nano) titanium dioxide (TiO2). It is available at http://www.cdc.gov/niosh/docs/2011-160/.
- *General Safe Practices for Working with Engineered Nanomaterials in Research Laboratories*, DHHS (NIOSH) Publication No. 2012–147 [NIOSH 2012b]. It is available at http://www.cdc.gov/niosh/ docs/2012-147/.
- NIOSH sponsored a supplement to the Journal of Occupational and Environmental Medicine, entitled Selected Papers from the Nanomaterials and Worker Health, Medical Surveillance, Exposure Registries and Epidemiologic Research Conference, July 21–23, 2010 [JOEM 53(6S):S1–S112].
- *Progress Toward Safe Nanotechnolog y in the Workplace: A Report from the NIOSH Nanotechnology Research Center*, DHHS (NIOSH) Publication No. 2010–104 [NIOSH 2009a]. This document reports on the status of nanotechnology research conducted by the NIOSH NTRC in 2007–2008. It is available at http://www. cdc.gov/niosh/docs/2010-104/.
- *Approaches to Safe Nanotechnology: Managing the Health and Safety Concerns Associated with Engineered Nanomaterials*, DHHS (NIOSH) Publication No. 2009–125 [NIOSH 2009c]. This document served a dual purpose of summarizing NIOSH's current thinking and interim recommendations and fostering the exchange information to ensure that no worker suffers material impairment of safety or health as nanotechnology develops. It is available at http://www.cdc. gov/niosh/docs/2009-125/.
- *Interim Guidance for Medical Screening and Hazard Surveillance for Workers Potentially Exposed to Engineered Nanoparticles*, DHHS (NIOSH) Publication No. 2009–116 [NIOSH 2009d]. This document

recommended prudent precautionary interim measures for reducing work-related exposures and assessing potential risk. It is available at http://www.cdc.gov/niosh/docs/2009-116/.

5.2. NIOSH Peer-Reviewed Publications

NIOSH NTRC scientists will continue to publish results of research as the data become available. More than 500 peer-reviewed articles were published between 2004 and 2012 that addressed scientific and technical issues in the field of nanotechnology. NIOSH will continue to evaluate the utility of its publications and their impact on the scientific, regulatory, and occupational health community.

5.3. Sponsored Conferences

The NIOSH NTRC will continue to partner with others in sponsoring and conducting conferences on nanotechnology. To date, NIOSH has co-sponsored 15 international nanomaterial health and safety meetings. NTRC staff members also have participated on several scientific and technical panels convened by government agencies, nongovernmental agencies, and professional associations.

5.4. Presentations

NTRC staff will continue to deliver presentations nationally and internationally concerning occupational safety and health issues associated with nanotechnology, including presentations at scientific conferences and to trade and professional associations. To date, NTRC staff members have given more than 650 presentations on nanomaterial health and safety topics.

6. Research to Practice (r2p)

Research to practice (r2p) involves the translation of research into products, practices, and usable information. The NIOSH NTRC nanotechnology strategic plan reflects the r2p vision to work with partners and stakeholders to translate research findings into NIOSH products (e.g., guidance documents, instrumentation, nanoparticle filtration methods, and exposure controls) that will be used to reduce or prevent worker injury and illness from nanotechnology. R2p also involves moving others to action. NIOSH will continue to work with national and international organizations to translate research to actions that will protect the nanomaterial workforce.

6.1. Capacity Building through Technical Assistance

The NIOSH NTRC is currently collaborating with a number of industries to develop appropriate engineering controls and effective administrative practices for the safe handling of nanomaterials. Specifically, the NTRC will continue to work with industry and labor in evaluating workplace exposures to ENMs and provide recommendations that will minimize worker exposures. NIOSH has several complimentary resources that reside in the Health Hazard Evaluation, Industrywide Study, and Engineering Control Technology Programs that can provide expertise in augmenting ongoing NTRC efforts of evaluating workplace nanoparticle exposures, related health effects, and risk management efforts. The NIOSH NTRC will continue to make recommendations and provide information based on its research and findings reported in the published scientific literature. The recommendations will pertain to all areas of risk management described in Section 4.1.

7. Intermediate Customers and Intermediate Outcomes

7.1. Federal Government Agencies

NIOSH will conduct and coordinate research with other agencies to foster the responsible development and safe use of nanotechnology, as identified by the NNI. Specifically, NIOSH will continue to collaborate with OSHA, EPA, NIST, NIEHS, and the Consumer Product Safety Commission (CPSC).

re EHS Research Categories and their Highest-Priority Research Needs (RNs)

Needs	Subordinate Research Needs
Measurement Infrastructure Research Needs ment tools to detect and identify engineered nanoscale materials elevant matrices and determine their physico-chemical properties ages of their life cycles. ement tools for determination of biological response and to enable azards and exposure for humans and the environment from omaterials and nanotechnology-based products throughout all ife cycles.	
p tools for of physico- perties of evant media and fe cycles of NEPs	• Physical dimensions and morphology: size, size distribution, characteristic dimensions, shape • Internal structure: atomic-molecular, core-shell • Surface and interfacial properties: surface charge, zeta potential, surface structure, elemental composition, surface-bound molecular coatings and conjugates, reactivity • Bulk composition: elemental or molecular composition, crystalline phase(s) • Dispersion properties: degree and state of dispersion • Mobility and other transport properties: diffusivity, transport in biological and environmental matrices
Develop ment tools for n and monitoring of in realistic exposure and conditions during cycles of ENMs and	• Sampling and collection of ENMs • Detecting the presence of ENMs • Quantity of ENMs-concentration based on surface area, mass, and number concentrations • Size and size distribution of ENMs • Spatial distribution of ENMs • Discriminating ENMs from ambient NMs such as combustion products and welding fumes • Discriminating multiple types of ENMs such as metals and metal oxides
#3. Develop asurement tools for luation of transfor- tions of ENMs in evant media and during e life cycles of ENMs and EPs	• Agglomeration and de-agglomeration • Dissolution and solubility • Absorption of natural organic matter and bioconstituents • Oxidation and reduction • Deposition of ENMs on surfaces

7.2. Standards Development Organizations

NIOSH actively participates in the development of national and international standards for promoting the health and safety of workers in the nanotechnology industries. The NIOSH NTRC participates in the American National Standards Institute (ANSI) Nanotechnology Standards Steering Panel, which coordinates the identification and development of critical standards in all areas of nanotechnology.

NIOSH NTRC scientists also participate in the ASTM International E56 Committee on Nanotechnology, which is developing an integrated family of standards. Committee E56.03 is addressing environmental and occupational safety and health.

NIOSH NTRC scientists will continue leading as chair and members of the U.S. Technical Advisory Group to the International Organization for Standardization (ISO) Technical Committee 229 on Nanotechnologies (ISO TC 229). Work on instrumentation-related standards will continue with committees of the International Electrotechnical Commission (IEC).

NIOSH NTRC scientists will also continue collaboration on the development of nanotechnology-related guidance with authoritative bodies including the National Council on Radiation Protection and Measurements.

7.3. Industry, Labor, and Academia

NIOSH is coordinating input from industry, labor, academics, and a wide range of government agencies in creating guidance for occupational health surveillance. Additionally, the NIOSH NTRC plans to coordinate input from those same groups of partners and stake-holders with the goal of developing standardized data systems for epidemiological research in workplaces producing and using nanomaterials. Through collaborations with industry, government, and academia, NIOSH [2009c] has developed a "best practices" document, *Approaches to Safe Nanotechnology: Managing the Health and Safety Concerns Associated with Engineered Nanomaterials* [DHHS (NIOSH) Publication No. 2009–125; http://www.cdc. gov/niosh/docs/2009-125/] and will continue to develop other communication materials on the safe handling of ENMs. NIOSH NTRC is continuing to work with industry to characterize occupational exposure to ENMs and implement effective risk-management practices to minimize worker exposure to ENMs.

7.4. Professional Organizations

NIOSH is collaborating with various professional organizations, including the American Industrial Hygiene Association (AIHA) and National Safety Council (NSC), to identify mutual efforts for developing new worker training programs.

7.5. Research Collaborations

NIOSH has established several national and international collaborations to advance research into the safe use of nanotechnology. The NIOSH NTRC participates in the National Nanotechnology Initiative (NNI) and has contributed to the nanotechnology environmental, health, and safety (EH&S) research strategy for the nation through the working group of Nanotechnology Environmental and Health Implications (NEHI). NIOSH's strategic research plan and activities have been developed to address the occupational safety and health issues in the NSET/NEHI plan [NNI 2011].

7.6. International Activities

NIOSH will continue to engage with a number of international entities at all levels—as principle investigator as well as in national, regional, and global organizations. At the national organization level, the NIOSH NTRC has been communicating and collaborating with the United Kingdom Institute of Occupational Medicine and the Health and Safety Laboratory; the Netherlands Organization for Applied Scientific Research (TNO); the French Agency for Food, Environmental and Occupational Health and Safety (ANSES); the Finnish Institute of Occupational Health; and the Australian Safety and Compensation Council (Safe Work Australia).

NIOSH is collaborating with the OECD to build cooperation, coordination, and communication between the United States and 30 OECD member countries, including the European Union (EU), and more than 180 non-member economies. NIOSH is also working with the UN World Health Organization (WHO); the UN International Labour Organization (ILO); the International Organization for Standardization (ISO); the International Electrotechnical Commission (IEC), the International Commission on Occupational Health (ICOH); and the International Council on

Key Research Needs	Subordinate Research Needs
RN#4. Develop measurement tools for evaluation of biological responses to ENMs and NEPs in relevant media and during the life cycles of ENMs and NEPs	• Adequacy of existing assays • New assays or high-throughput, high content assays • Correlation of biological responses with physico-chemical properties • Surface reactivity at the interfaces between ENMs and biological receptors • Biomarkers of toxicological response
RN#5. Develop measurement tools for evaluation of release mechanisms of ENMs from NEPs in relevant media and during the life cycles of NEPs	• Release by fire, combustion, and incineration • Release by mechanical degradation, such as abrasion, deformation, and impact • Release by dissolution of matrix material • Release by chemical reactions of the matrix material • Release by photo-induced degradation of the matrix material • Release by consumer interactions, such as spraying, mouthing, and swallowing • Release by interactions with biological organisms in the environment
2. Human Exposure Assessment Research Needs Identify potential sources, characterize the exposure scenarios, and quantify actual exposures of workers, the general public and consumers to nanomaterials. Characterize and identify the health outcomes among exposed populations in conjunction with information about the control strategies used and exposures to determine practices that result in safe levels of exposure.	
RN#1. Understand processes and factors that determine exposures to nanomaterials	• Conduct studies to understand processes and factors that determine exposure to engineered nanomaterials • Develop exposure classifications of nanomaterials and processes • Develop internationally harmonized and validated protocols for exposure surveys, sample collection and analysis, and reporting through existing and newly created international frameworks • Develop comprehension predictive models for exposures to a broad range of engineered nanomaterials and processes • Characterize process- and task-specific exposure scenarios in the workplace

Table D–1. (Continued)

Key Research Needs	Subordinate Research Needs
RN#2. Identify population groups exposed to engineered nanomaterials	• Systematically collect and analyze information about nanomaterial manufacture, processing, and direct use in commercial and consumer products over time to discern geographic areas where engineered nanomaterials may be emitted into the environment, consumed in the form of ingredients of products, and/or disposed of in solid waste, wastewater, etc. • Conduct population-based surveys to obtain information on use patterns for consumer products • Identify potential subpopulations that are more susceptible to exposure to engineered nanomaterials than others • Develop quantitative assessment methods appropriate for target population groups and conduct assessments of those population groups most likely to be exposed to engineered nanomaterials
RN#3. Characterize individual exposures to nanomaterials	• Expand currently available exposure assessment techniques to facilitate more accurate exposure assessment for engineered nanomaterials at benchmark concentration levels using feasible methods • Develop new tools through national and international surveys to support effective exposure characterization of individuals • Characterize and detect nanomaterials in biological matrices and conduct studies to understand transformations of nanomaterials during transport in the environment and in human bodies • Conduct studies to examine emissions and human contact during normal use and after wear and tear have degraded a product, as well as during repeated exposures • Develop engineered nanomaterials exposure assessment models based on identified critical exposure descriptions • Develop databases to contain the collected data and information

Key Research Needs	Subordinate Research Needs
RN#4. Conduct health surveillance of exposed populations	• Establish a program for the epidemiological investigations of physician case reports and reports of suspicious pattern of adverse events • Establish exposure registry and medical surveillance programs for workers • Analyze injury and illness reporting in existing programs
3. Human Health Research Needs Understand the relationship of physico-chemical properties of engineered nanoscale materials to *in vivo* physico-chemical properties and biological response. Develop high-confidence predictive models of *in vivo* biological responses and casual physico-chemical properties of ENMs.	
RN#1. Identify or develop appropriate, reliable, and reproducible *in vitro* and *in vivo* assays and models to predict *in vivo* human responses to ENMs	• Establish a system to develop and apply reliable and reproducible *in vitro* and *in vivo* test methods • Evaluate the degree to which an *in vitro* response correlates with an *in vivo* response • Evaluate the degree to which *in vitro* and *in vivo* models predict human response • Translate structure-activity relationship and other research data into computational models to predict toxicity *in silico*
RN#2. Quantify and characterize ENMs in exposure matrices and biological matrices	• Determine critical ENM measurands in biological and environmental matrices and ensure the development of tools to measure ENMs in appropriate matrices as needed • Determine matrix and/or weathering effects that may alter the physico-chemical characteristics of the ENM measurands • Identify key factors that may influence the detection of each measurand in a particular matrix (e.g., sample preparation, detection method, storage, temperature, solvents/solutions) • Characterize and quantify exposure for all exposure routes using *in vivo* models to identify the most likely routes of human exposure • Identify biomarkers of exposure and analytical methods for their determination
RN#3. Understand the relationship between the physico-chemical properties of ENMs and their transport, distribution, metabolism, excretion, and body burden in the human body	• Characterize ENM physico-chemical properties and link to mechanisms of transport and distribution in the human body • Understand the relationship of the physico-chemical properties of ENMs to the mechanisms of sequestration in and translocation of ENMs out of the exposure organ and secondary organs, and to routes of excretion from the human body

Table D–1. (Continued)

Key Research Needs	Subordinate Research Needs
	• Determine the metabolism or biological transformation of ENMs in the human body
RN#4. Understand the relationship between the physico-chemical properties of ENMs and uptake through the human port-of-entry tissues	• Characterize ENMs at and in port-of-entry tissues, including nontraditional routes of entry such as the ear and eye, and identify mechanisms of ENM uptake into tissues • Determine the relationship of ENM physico-chemical properties to deposition and uptake under acute exposure conditions and under chronic exposure conditions • Translate data on ENM properties and uptake to knowledge that may be used to intentionally redesign ENMs for optimum human and environmental safety and product efficacy
RN#5. Determine the modes of action underlying the human biological response to ENMs at the molecular, cellular, tissue, organ, and whole body levels	• Determine the dose response and time course of biological responses at the primary site of exposure and at distal organs following ENM exposure • Understand the mechanisms and molecular pathway(s) associated with ENM biology within cellular, organ, and whole organism systems • Link mechanisms of response with ENM physico-chemical properties and employ this information in the design and development of future ENMs • Develop translational alternative *in vitro* testing methods for the rapid screening of future ENMs based on mechanism(s) of response that are predictive of *in vivo* biological responses
RN#6. Determine the extent to which life stage and/or susceptibility factors modulate health effects associated with exposure to ENMs and nanotechnology-enabled products and applications	• Determine the effect of life stage and/or gender on biological response to ENMs • Establish the role of genetic and epigenetic susceptibility on the biological response to ENMs in the context of life stage and/or susceptibility factors • Understand mechanistically the influence of preexisting disease on the biological response to ENMs in the context of life stage and other susceptibility factors • Identify exposure conditions that make susceptible individuals more vulnerable to the health effects associated with ENMs and nanotechnology-enabled applications

Key Research Needs	Subordinate Research Needs
	• Establish a database that contains published, peer-reviewed literature, occupational and consumer reports, and toxicological profiles that describe altered responses to ENMs and nanotechnology-enabled applications in susceptible animal models or individuals following exposure
4. Environment Research Needs Understand the environmental fate, exposure, and ecological effects of engineered nanomaterials, with priority placed on materials with highest potential for release, exposure, and/or hazard to the environment.	
RN#1. Understand environmental exposures through the identification of principal sources of exposure and exposure routes	• Manufacturing processes and product incorporation • Life cycle of technology and exposures subsequent to product manufacturing • Analytical approaches to measure temporal changes in nanoparticle properties throughout the life cycle • Models to estimate releases • Environmental receptors for exposure assessment
RN#2. Determine factors affecting the environmental transport of nanomaterials	• Determine key physico-chemical properties affecting transport • Determine key transport and fate processes relevant to environmental media • Develop new tools and adaptation of current predictive tools to accommodate unique properties of nanomaterials
RN#3. Understand the transformation of nanomaterials under different environmental conditions	• Identify and evaluate nanomaterial properties and transformation processes that will reduce environmental persistence, toxicity, and production of toxic products • Determine the rate of aggregation and long-term stability of agglomeration/aggregation and the long-term stability of these aggregates and agglomerates • Develop tools to predict the transformations or degradability of nanomaterials
RN#4. Understand the effects of engineered nanomaterials on individuals of a species and the applicability of testing schemes to measure effects	• Test protocols • Dose-response characterization • Uptake/elimination kinetics, tissue/organ distribution • Mode/mechanism of action, predictive tools • Tiered testing schemes/environmental realism

Table D–1. (Continued)

Key Research Needs	Subordinate Research Needs
RN#5. Evaluate the effects of engineered nanomaterials at the population, community, and ecosystem levels	• Population • Community • Other ecosystem-level effects • Predictive tools for population-, community-, and ecosystem-level effects
5. Risk Assessment and Risk Management Methods Research Needs Increase available information for better decision making in assessing and managing risks from nanomaterials, including using comparative risk assessment and decision analysis; life cycle considerations; and additional perspectives such as ELSI considerations, stakeholders' values, and additional decision makers' considerations.	
RN#1. Incorporate relevant risk characterization information, hazard identification, exposure science, and risk modeling and methods into the safety evaluation of nanomaterials	• Characterization, fate, and release of nanoparticles throughout the life cycles of nanotechnology-enabled products • Development of predictive models on accumulation, migration, and release of nanoparticles throughout the life cycles of nanotechnology-enabled products • Safety of nanoparticles throughout the life cycles of the nanotechnology-enabled products • Comprehensive and predictive models to assess the potential risks of nanoparticles during the manufacturing and life cycle of nanoproducts, with inputs from human and environment exposures and on nanomaterial properties
RN#2. Understand, characterize, and control workplace exposures to nanomaterials	• Dissemination and implementation of effective techniques and protocols to measure exposures in the workplace • Identification and demonstration of effective containment and control technologies including for accidents and spills • Development of an effective industry surveillance system • Design and deployment of a prospective epidemiological framework relevant to exposure science • Systematic approaches for occupational risk modeling
RN#3. Integrate life cycle considerations into risk assessment	• Establishment of nanotechnology-specific taxonomy for life cycle stages

Key Research Needs	Subordinate Research Needs
	• Integration of risk assessment, life cycle analyses, and decision-making approaches into regulatory decision making processes • Application of adaptive management tools based on monitoring/implementation to evaluate life cycle analysis implementation • Development of case studies, e.g., green chemistry, nanomaterials selection, nanomaterials acquisition process, illustrating application of these risk management methods
RN#4. Integrate risk assessment into decision-making frameworks for risk management	• Development of comparative risk assessment and formal decision-analytical methods as opposed to "absolute" risk assessment strategies • Application of formal decision-analytical methods to prioritize risk management alternatives • Use of gap analyses and value of information analysis to identify research needs • Integration of stakeholder values and risk perceptions into risk management processes • Application of integrated decision framework through case studies in risk management decision making
RN#5. Integrate and standardize risk communication within the risk management framework	• Development and use of standardized terminology in risk communications • Early information-sharing on hazards and risk among Federal agencies • Development of appropriate risk communication approaches for agency-specific needs
Informatics and Modeling Research Need	
RN#1. Develop computational models of ENM structure- property-activity relationships to support the design and development of ENM with maximum benefit and minimum risk to humans and the environment	• Validate the predictive capability of *in vitro* and *in vivo* assays and employ that subset of assays in data generation to establish computational models to predict ENM behavior in humans and the environment • Establish a standard set of physical and chemical characterization parameters, dose metrics, and biological response metrics • Design and establish structures and ontologies for methods development, data capture, sharing, and analysis • Evaluate and adapt as necessary existing computational models by beginning with existing models for exposure and dosimetry and using data generated from validated assays

Table D–1. (Continued)

Key Research Needs	Subordinate Research Needs
	• Use ENM exposure and dosimetry models to develop ENM structure-activity models to predict ENM behavior in humans and the environment • Establish training sets and beta test sites to refine and validate ENM structure-activity models • Disseminate ENM structure-activity models through publicly accessible nanotechnology websites

Appendix B: Timeline for NIOSH Nanotechnology Research

Summary of Proposed Research Projects, FY2013–FY2014

Critical research area	Projects
Toxicity and Internal Dose	Evaluate long-term fibrogenic and carcinogenic potential of carbon nanotubes and carbon nanofibers. Determine immune and cardiovascular responses to pulmonary exposure to CNT, TiO_2, and nanosilver. Evaluate the pulmonary effects of nanosilver.
Measurement Methods	Develop updated methods and approaches to assess occupational exposures to nanomaterials. Utilize reference materials and technically sound protocols to improve measurement quality. Develop techniques to characterize nanoparticles, using advanced microscopy methods. Study the performance of direct reading instruments for assessing emissions and exposure to nanoparticles in the workplace.
Exposure Assessment	Complete evaluation of workplace exposures to emerging, second-generation, and third-generation ENMs and their potential routes of exposure. Complete evaluation of size, concentration, and morphology of ENMs emitted by various processes.
Epidemiology and Surveillance	Evaluate occupational exposures among carbonaceous nanomaterial workers.

Critical research area	Projects
	Complete field work for a cross-sectional epidemiologic survey among workers exposed to carbonaceous nanomaterials. Collect data on specific biomarkers as markers of exposure or early health endpoints in workers exposed to carbon nanotubes or carbon nanofibers.
Risk Assessment	Develop a strategy and criteria to utilize the best available scientific data to develop hazard- and risk-based occupational exposure limits and other risk management guidance. Identify nanomaterial categories by physicochemical properties and biological mode of action, and perform case study evaluations for nanomaterials within each major category. Analyze toxicity data, including identifying the adverse health endpoint(s) and performing dose-response analyses and/or comparative potency analyses. Contribute to the development and evaluation of dosimetry models and methods to assess internal dose of nanomaterials in workers.
Engineering Controls and PPE	Conduct field studies to assess and improve engineering control strategies in the workplace. Evaluate effectiveness of commercially available engineering controls in the laboratory. Finalize development of a standardized aerosol test method using magnetic passive aerosol samplers for testing of protective clothing material. Expand particulate penetration project to include evaluation of different types of fabrics, use of electrostatic charges, and bellows effects. Evaluate balance between comfort and protection of protective clothing and gloves with ENMs.
Fire and Explosion Safety	Evaluate nano metals and other ENMs for fire and explosion potential. Evaluate a variety of ENMs for electromagnetic hazard potential. Correlate dustiness of ENMs with various physical and chemical safety hazards. Evaluate current DOT and NFPA guidance for adequacy with ENMs. Provide recommendations to first responders about safety issues of ENMs.
Recommendations and Guidance	Continue conducting r2p activities (ongoing), such as developing brochures and fact sheets and updating the topic page.

(Continued)

Critical research area	Projects
	Develop a small business guide for nanomaterial producers and users.
Applications	Further develop and refine end-of-service indicators using nanomaterials.
Global Activities	Strengthen participation with globally recognized organizations ICON, ISO, OECD, and UN (WHO, ILO). Participate in the US/EU nanomaterial working groups hosted by the NNI.

Summary of Proposed Research Projects for FY2015–FY2016

Critical research area	Projects
Toxicity and Internal Dose	Identify biomarkers of response to ENMs. Develop predictive algorithms for relationships between nanoparticle properties and bioactivity. Determine the relevance of in vitro and in vivo screening tests. Evaluate the pulmonary and systemic effects of graphene and nanocellulose.
Measurement Methods	Publish updated exposure measurement methods. Assist with round-robin evaluations of ENM reference materials.
Exposure Assessment	Publish exposure results. Provide guidance on exposure assessment methods and sampling protocols.
Epidemiology and Surveillance	Analyze biomarker samples from cross-sectional epidemiologic study of carbon nanotube and nanofiber workers. Analyze data (medical and biomarker with exposure) from cross-sectional epidemiologic study of carbon nanotube and nanofiber workers. Begin collecting data for exposure registry and prospective epidemiologic study among carbon nanotube–exposed and carbon nanofiber–exposed workers. Assess feasibility of epidemiologic study of health outcomes among workers exposed to nanomaterials other than carbonaceous nanomaterials.

Critical research area	Projects
Risk Assessment	Extend evaluations of the hazard and risk of workplace exposure to new nanomaterials. Utilize the latest data from toxicity studies, workplace exposure measurements, and/or biomonitoring data to identify potential hazards in the respiratory tract and/or other organs. Contribute to the refinement and validation of dosimetry models for nanomaterials. Contribute to the evaluation and development or updating of recommended exposure limits and other risk management guidance.
Engineering Controls and PPE	Continue to assess engineering controls in the field and update recommendations for controlling exposure. Continue to evaluate commercially available engineering controls in the lab and publish test results. Evaluate respirator performance with ENMs in laboratory and workplace settings. Develop a passive method for measuring particle penetration through protective clothing materials.
Fire & Explosion Safety	Continue evaluation of various ENMs for fire and explosion potential. Evaluate a variety of ENMs for electromagnetic hazard potential. Correlate dustiness of ENMs with various physical and chemical safety hazards. Publish findings in journals and provide summaries to DOT and NFPA.
Recommendations and Guidance	Update Approaches to Safe Nanotechnology Develop hazard bands and RELs for high-volume commercially available ENMs.
Applications	Evaluate ENM-enabled PPE.
Global Activities	Lead or co-lead committees with globally recognized organizations: ICON, ISO, OECD, and UN (WHO, ILO).

NOTE: DOT, Department of Transportation; ENMs, engineered nanomaterials; ICON, International Council on Nanotechnology; ILO, International Labour Organization; ISO, International Organization for Standardization; NFPA, National Fire Protection Association; NNI, National Nanotechnology Initiative; OECD, Organization for Economic Cooperation and Development; PPE, personal protective equipment; UN, United Nations; WHO, World Health Organization.

References

Gibbs, L. (2006). Presentation at the annual meeting of the Campus Safety Health and Environmental Management Association (CSHE-MA), Anaheim, CA, July 15–19.

Howard, J. & Murashov, V. (2009). National nanotechnology partnership to protect workers. *J Nanopart Res*, *11*, 1673–1683.

ISO. (2008). Nanotechnologies, terminology and definitions for nano-objects—nanoparticle, nanofibre and nanoplate. ISO/TS 27687. Geneva, Switzerland, International Organization for Standardization.

Murashov, V., Schulte, P. A. & Howard, J. (2012). Progression of occupational risk management with advances in nanomaterials. *J Occup Environ Health*, *9*, D12–D22.

Nasterlack, M., Zober, A. & Oberlinner, C. (2008). Considerations in occupational medical surveillance of employees handling nanoparticles. *Int Arch Occup Environ Health*, *81*, 721–726.

NIOSH. (2007). Progress toward safe nanotechnology in the workplace. Cincinnati, OH: U.S. Department of Health and Human Services, Centers for Disease Control and Prevention, National Institute for Occupational Safety and Health, DHHS (NIOSH) Publication No. 2007–123.

NIOSH. (2009a). Progress toward safe nanotechnology in the workplace: a report from the NIOSH Nanotechnology Research Center. Cincinnati, OH: U.S. Department of Health and Human Services, Centers for Disease Control and Prevention, National Institute for Occupational Safety and Health, DHHS (NIOSH) Publication No. 2010–104.

NIOSH. (2009b). Strategic plan for NIOSH nanotechnology research and guidance. Cincinnati, OH: U.S. Department of Health and Human Services, Centers for Disease Control and Prevention, National Institute for Occupational Safety and Health, DHHS (NIOSH) Publication No. 2010–105.

NIOSH. (2009c). Approaches to safe nanotechnology: managing the health and safety concerns associated with engineered nanomaterials. Cincinnati, OH: U.S. Department of Health and Human Services, Centers for Disease Control and Prevention, National Institute for Occupational Safety and Health, DHHS (NIOSH) Publication No. 2009–125.

NIOSH. (2009d). Interim guidance for medical screening and hazard surveillance for workers potentially exposed to engineered nanoparticles. Cincinnati, OH: U.S. Department of Health and Human Services, Centers for Disease Control and Prevention, National Institute for Occupational Safety and Health, DHHS (NIOSH) Publication No. 2009–116.

NIOSH. (2011). Current intelligence bulletin 63: occupational exposure to titanium dioxide. Cincinnati, OH: U.S. Department of Health and Human Services, Centers for Disease Control and Prevention, National Institute for Occupational Safety and Health, DHHS (NIOSH) Publication No. 2011–160.

NIOSH. (2012a). Filling the knowledge gaps: a report from the NIOSH Nanotechnology Research Center, from inception in 2004 through 2011. Cincinnati, OH: U.S. Department of Health and Human Services, Centers for Disease Control and Prevention, National Institute for Occupational Safety and Health, DHHS (NIOSH) Publication No. 2013–101.

NIOSH. (2012b). General safe practices for working with engineered nanomaterials in research laboratories. Cincinnati, OH: U.S. Department of Health and Human Services, Centers for Disease Control and Prevention, National Institute for Occupational Safety and Health, DHHS (NIOSH) Publication No. 2012–147.

NNI. (2001). National Nanotechnology Initiative: interagency agreement. Washington, DC: United States Executive Office of the President, Office of Science and Technology Policy [http://www.nano.gov/html/about/home_about. Html].

NNI. (2011). National Nanotechnology Initiative: environmental, health, and safety research strategy. Washington, DC: United States Executive Office of the President, Office of Science and Technology Policy, Nanotechnology Environmental and Health Implications (NEHI) Working Group.

NRC. (2009). Science and decisions: advancing risk assessment. Committee on improving risk analysis approaches used by the U.S. EPA, Board on Environmental Studies and Toxicology, Division on Earth and Life Studies, National Research Council of the National Academies.

NSF. (2011). Nanotechnology research directions for societal needs in 2020: retrospective and outlook summary. National Science Foundation. In: Roco M, Mirkin C, Hersan M, eds. Science Policy Reports. New York: Springer (DOI: 10.1007/978-94-007-1168-6_5).

Schulte, P., Geraci, C., Zumwalde, R., Hoover, M., Castranova, V., Kuempel, E., Murashov, V., Vainio, H. & Savolainen, K. (2008a). Sharpening the focus on occupational safety and health of nanotechnology. *Scan J Work Environ Health*, *34*(6), 471–478.

Schulte, P., Geraci, C., Zumwalde, R., Hoover, M. & Kuempel, E. (2008b). Occupational risk management of engineered nanoparticles. *J Occup Environ Hyg*, *5*, 239–249.

Schulte, P. A. & Salamanca-Buentello, F. (2007). Ethical and scientific issues of nanotechnology in the workplace. *Environ Health Perspect*, *115*, 5–12.

Schulte, P. A., Schubauer-Berigan, M. K., May-weather, C., Geraci, C. L., Zumwalde, R. & McKernan, J. L. (2009). Issues in the development of epidemiologic studies of workers exposed to engineered nanoparticles. *J Occup Environ Med*, *51*(3), 323–335.

In: Nanotechnology Workforce Protection
Editor: Suzanne G. Rice
ISBN: 978-1-63117-641-8

Chapter 2

GENERAL SAFE PRACTICES FOR WORKING WITH ENGINEERED NANOMATERIALS IN RESEARCH LABORATORIES*

National Institute for Occupational Safety and Health

FOREWORD

The National Institute for Occupational Safety and Health (NIOSH) is pleased to present *General Safe Practices for Working with Engineered Nanomaterials in Research Laboratories.* Engineered nanomaterial applications are rapidly expanding throughout the United States and worldwide. The research community is at the front line of creating these new nanomaterials, testing their usefulness in a variety of applications, and determining their toxicological and environmental impacts.

With the publication of this document, NIOSH hopes to raise awareness of the occupational safety and health practices that should be followed during the synthesis, characterization, and experimentation with engineered nanomaterials in a laboratory setting. The document contains recommendations on engineering controls and safe practices for handling

---

* This is an edited, reformatted and augmented version of DHHS (NIOSH) Publication No. 2012–147, dated May 2012; released by U.S. Department of Health and Human Ser-vices, Centers for Disease Control and Prevention, National Institute for Occupational Safety and Health.

engineered nanomaterials in laboratories and some pilot scale operations. This guidance was designed to be used in tandem with well-established practices and the laboratory's chemical hygiene plan. As our knowledge of nanotechnology increases, so too will our efforts to provide additional guidance materials for working safely with engineered nanomaterials.

/s
John Howard, M.D.
Director, National Institute for
Occupational Safety and Health
Centers for Disease Control and Prevention

EXECUTIVE SUMMARY

Nanotechnology, the manipulation of matter at a nanometer scale to produce new materials, structures, and devices having new properties, may revolutionize life in the future. It has the potential to impact medicine through improved disease diagnosis and treatment technologies and to impact manufacturing by creating smaller, lighter, stronger, and more efficient products. Nanotechnology could potentially decrease the impact of pollution by improving methods for water purification or energy conservation. Although engineered nanomaterials present seemingly limitless possibilities, they bring with them new challenges for identifying and controlling potential safety and health risks to workers. Of particular concern is the growing body of evidence that occupational exposure to some engineered nanomaterials can cause adverse health effects.

As with any new technology or new material, the earliest exposures will likely occur for those workers conducting discovery research in laboratories or developing production processes in pilot plants. The research community is at the front line of creating new nanomaterials, testing their usefulness in a variety of applications and determining their toxicological and environmental impacts. Researchers handling engineered nanomaterials in laboratories should perform that work in a manner that protects their safety and health. This guidance document provides the best information currently available on engineering controls and safe work practices to be followed when working with engineered nanomaterials in research laboratories.

Risk Management

Risk management is an integral part of occupational health and safety. Potential exposures to nanomaterials can be controlled in research laboratories through a flexible and adaptive risk management program. An effective program provides the framework to anticipate the emergence of this technology into laboratory settings, recognize the potential hazards, evaluate the exposure to the nanomaterial, develop controls to prevent or minimize exposure, and confirm the effectiveness of those controls.

Hazard Identification

Experimental animal studies indicate that potentially adverse health effects may result from exposure to nanomaterials. Experimental studies in rodents and cell cultures have shown that the toxicity of ultrafine particles or nanoparticles is greater than the toxicity of the same mass of larger particles of similar chemical composition.

Research demonstrates that inhalation is a significant route of exposure for nanomaterials. Evidence from animal studies indicates that inhaled nanoparticles may deposit deep in lung tissue, possibly interfering with lung function. It is also theorized that nanoparticles may enter the bloodstream through the lungs and transfer to other organs. Dermal exposure and subsequent penetration of nanomaterials may cause local or systemic effects. Ingestion is a third potential route of exposure. Little is known about the possible adverse effects of ingestion of nanomaterials, although some evidence suggests that nanosized particles can be transferred across the intestinal wall.

Exposure Assessment

Exposure assessment is a key element of an effective risk management program. The exposure assessment should identify tasks that contribute to nanomaterial exposure and the workers conducting those tasks. An inventory of tasks should be developed that includes information on the duration and frequency of tasks that may result in exposure, along with the quantity of the material being handled, dustiness of the nanomaterial, and its physical form. A thorough understanding of the exposure potential will guide exposure assessment measurements, which will help determine the type of controls required for exposure mitigation.

Exposure Control

Exposure control is the use of a set of tools or strategies for decreasing or eliminating worker exposure to a particular agent. Exposure control consists of a standardized hierarchy to include (in priority order): elimination, substitution, isolation, engineering controls, administrative controls, or if no other option is available, personal protective equipment (PPE).

Substitution or elimination is not often feasible for workers performing research with nanomaterials; however, it may be possible to change some aspects of the physical form of the nanomaterial or the process in a way that reduces nanomaterial release.

Isolation includes the physical separation and containment of a process or piece of equipment, either by placing it in an area separate from the worker or by putting it within an enclosure that contains any nanomaterials that might be released.

Engineering controls include any physical change to the process that reduces emissions or exposure to the material being contained or controlled. Ventilation is a form of engineering control that can be used to reduce occupational exposures to airborne particulates. General exhaust ventilation (GEV), also known as dilution ventilation, permits the release of the contaminant into the workplace air and then dilutes the concentration to an acceptable level. GEV alone is not an appropriate control for engineered nanomaterials or any other uncharacterized new chemical entity. Local exhaust ventilation (LEV), such as the standard laboratory chemical hood (formerly known as a laboratory fume hood), captures emissions at the source and thereby removes contaminants from the immediate occupational environment. Using selected forms of LEV properly is appropriate for control of engineered nanomaterials.

Administrative controls can limit workers' exposures through techniques such as using job-rotation schedules that reduce the time an individual is exposed to a substance. Administrative controls may consist of standard operating procedures, general or specialized housekeeping procedures, spill prevention and control, and proper labeling and storage of nanomaterials. Employee training on the appropriate use and handling of nanomaterials is also an important administrative function.

PPE creates a barrier between the worker and nanomaterials in order to reduce exposures. PPE may include laboratory coats, impervious clothing, closed-toe shoes, long pants, safety glasses, face shields, impervious gloves, and respirators.

Other Considerations

Control verification or confirmation is essential to ensure that the implemented tools or strategies are performing as specified. Control verification can be performed with traditional industrial hygiene sampling methods, including area sampling, personal sampling, and real-time measurements. Control verification may also be achieved by monitoring the performance parameters of the control device to ensure that design and performance criteria are met.

Other important considerations for effective risk management of nanomaterial exposure include fire and explosion control. Some studies indicate that nanomaterials may be more prone to explosion and combustion than an equivalent mass concentration of larger particles.

Occupational health surveillance is used to identify possible injuries and illnesses and is recommended as a key element in an effective risk management program. Basic medical screening is prudent and should be conducted under the oversight of a qualified healthcare professional.

ABBREVIATIONS

ACGIH	American Conference of Governmental Industrial Hygienist
AIHA	American Industrial Hygiene Association
ANSI	American National Standards Institute
ASHRAE	American Society of Heating, Refrigeration, and Air Conditioning Engineers
BSC	biological safety cabinet
CFR	Code of Federal Regulations
CHN	Center for High-rate Nanomanufacturing
CNT	carbon nanotube
DNA	deoxyribonucleic acid
FLAR	flame aerosol reactor
FSP	flame spray pyrolysis

GEV	general exhaust ventilation
HEPA	high-efficiency particulate air
HVAC	heating, ventilation, and air conditioning
ICRP	International Commission on Radiological Protection
ISO	International Organization for Standardization
LEV	local exhaust ventilation
LLNL	Lawrence Livermore National Laboratory
µm	micrometer
MWCNT	multi-walled carbon nanotube
nm	nanometer
NAS	National Academy of Sciences
NIOSH	National Institute for Occupational Safety and Health
NTRC	Nanotechnology Research Center
OPC	optical particle counter
OECD	Organization for Economic Co-operation and Development
OSHA	Occupational Safety and Health Administration
PEL	permissible exposure limit
PtD	Prevention through Design
REL	recommended exposure limit
RL	risk level
SWCNT	single-walled carbon nanotube

1. Introduction

According to The International Organization for Standardization Technical Committee 229 (Nanotechnologies) (ISO/TS 27687:2008), a nano-object is a material with one, two, or three external dimensions in the 1- to 100-nm size range. Nano-objects are frequently incorporated into a larger matrix known as a nanomaterial. Nanoparticles are a specific type of nano-object, with all three external dimensions at the nanoscale. An additional term, ultrafine particles, is used to describe nanometer-diameter particles that have not been intentionally produced but are the incidental products of processes (NIOSH 2009a). For purposes of this document, the term *nanomaterial* is used to describe engineered nano-objects, including engineered nanoparticles.

Nanomaterials are increasingly being used in optoelectronic, electronic, magnetic, medical imaging, drug delivery, cosmetic, catalytic, and other applications. Although nano-materials present seemingly limitless possibilities, they bring with them new challenges to understanding, predicting, and managing potential safety and health risks to workers. Exposures to nanomaterials can involve a wide range of nanomaterial sizes, shapes, functionalities, concentrations, chemical compositions, and exposure frequencies or durations. Researchers working with engineered nanomaterials have the potential to be exposed through a variety of sources and processes, including leaks from equipment used in the synthesis of nanomaterials, manipulating dry nanopowders, sonicating liquid suspensions, or mechanically disrupting materials containing or coated with nano-materials (Aitken et al. 2004; Johnson et al. 2010). A growing body of evidence indicates that exposure to some of these engineered nanomaterials can cause adverse health effects. Based on this preliminary toxicological data, prudent practice dictates controlling occupational exposure to nanomaterials.

2. Scope

As with any new technology, the earliest exposures will likely occur among those workers conducting research in laboratories and pilot plants. Researchers handling engineered nanomaterials in laboratories and pilot scale operations should perform that work in a manner that is protective of their safety and health. Although incidental nanoparticles (also known as ultrafine particles) exist in nature, the focus of this document is to provide guidance on the safe handling of purposely designed, engineered nanomaterials in research laboratories. The information may also be applicable in some pilot-scale facilities.

Research laboratories include any facility performing basic or applied research involving nanomaterials. Nanomaterial research laboratories may be housed at universities, government agencies, and private companies. Research laboratories may produce their own nanomaterials, work with nanomaterials produced by others, or some combination of both. Laboratory-scale production typically consists of relatively small amounts of nanomaterial, ranging from a few milligrams for highly sophisticated materials such as quantum dots to a few kilograms for less-sophisticated materials such as metal oxides. Laboratories conducting applied research may also produce materials on a pilot scale, which typically increases material volumes by a factor of 10 or more. Pilot-scale equipment is generally similar to industrial-scale processes, but it produces much smaller quantities of nanomaterial.

3. Risk Management

Exposures to engineered nanomaterials can be controlled in the research laboratory by a comprehensive risk management program that includes task hazard/risk analysis, engineering controls, administrative controls, and use of PPE. Implementing an effective program should address the following elements of hazard surveillance.

Hazard Identification: Is there reason to believe that the nanomaterial of interest could be harmful?

Exposure Assessment: Is there potential for exposure to the nanomaterial or other chemical or physical hazards?

Exposure Control: What procedures are in place or should be developed to minimize or eliminate worker exposure(s)?

The answers to these questions will help to formulate a program that includes the following:

- A written health and safety policy covering all types of chemical and physical hazards in the workplace, in accordance with the U.S. regulatory requirement 29 CFR 1910.1450, the Occupational Safety and Health Administration's (OSHA's) laboratory standard, including development of a Chemical Hygiene Plan.
- A clear delineation of roles and responsibilities for everyone involved in laboratory or pilot plant research.
- Effective procedures for documentation, communication, and employee training.
- Incorporation of input from safety professionals, industrial hygienists, and occupational health professionals, as appropriate.

Figure 1 illustrates components of an overall health and safety program that includes nanomaterial risk management (Schulte et al. 2008a). Additional guidance on prudent practices in the laboratory (NRC 2011) can be obtained from the National Academy of Sciences (NAS).

Figure 1. Components of an overall health and safety program. Modified from Schulte et al. (2008).

4. Hazard Identification

The unique properties of materials at the nanoscale have raised concerns regarding health effects that might result from occupational exposures. The toxicity of a nanomaterial will be a function of its substance-specific toxicity, as influenced by physicochemical characteristics (including those unique to the nanoscale form of the substance) and contaminants (Trout and Schulte 2010).

Results of studies in which animals and humans were exposed to ultrafine or other respirable particles provide a basis of concern for possible adverse health effects due to engineered nanomaterial exposures. Experimental studies in rodents and cell cultures have shown that the toxicity of ultrafine or nanoparticles is greater than that of the same mass of larger particles of similar chemical composition (Oberdörster et al. 1992; Oberdörster et al. 1994; Lison et al. 1997; Tran et al. 1999; Tran et al. 2000; Brown et al. 2001; Barlow et al. 2005; Duffin et al. 2007). In addition to particle size and surface area, other particle characteristics may influence toxicity, including surface functional groups or coatings, solubility, shape, and the ability to generate reactive oxygen species (Duffin et al. 2002; Maynard and Kuempel 2005; Oberdörster et al. 2005; Donaldson et al. 2006).

Several articles have investigated the toxicity of carbon nanotubes (CNTs) in experimental animal studies (Lam et al. 2004; Shvedova et al. 2005; Donaldson et al. 2006; Lam et al. 2006; Kisin et al. 2007; Li et al. 2007; Kane and Hurt 2008; Miyawaki et al. 2008; Poland et al. 2008; Shvedova et al. 2008; Erdely et al. 2009; Ma-Hock et al. 2009; Shvedova et al. 2009; Pauluhn 2010). The results from these studies indicate potential respiratory health risks from exposure to CNTs, including granulomatous pneumonia and fibrosis. Evidence also indicates that when multi-walled carbon nanotubes (MWCNTs) are administered intraperitoneally to mice, the MWCNTs have asbestos-like pathogenicity (Poland et al. 2008; Takagi et al. 2008). Although a causal link has not been established, there is concern about possible cancer hazards in addition to potential for fibrosis/nonmalignant respiratory disease.

Additional studies have investigated the DNA damage caused by nanosized metals and metal oxides (Karlsson et al. 2009; Singh et al. 2009). Although it cannot be concluded that metal oxide nanoparticles are always more toxic than their micrometer counterparts, nanosized copper oxide (CuO) was found to be much more toxic than micrometer-sized CuO (Karlsson et al. 2009).

Inhalation is considered the primary route of potential exposure in the nanomaterial workplace. Evidence indicates discrete nanoparticles are deposited in the lungs to a greater extent than larger respirable particles (ICRP 1994). Some nanoparticles are thought to enter the bloodstream from the lungs and then transfer to other organs (Takenaka et al. 2001; Nemmar et al. 2002; Oberdörster et al. 2002; Geiser et al. 2005). It is further postulated that some nanomaterials may move from the nose to the brain though the blood-brain barrier (Oberdörster et al. 2004; Elder et al. 2006).

Dermal exposure to nanomaterials is also a potential exposure pathway. Possible harmful effects may occur locally, or the substances may be absorbed through the skin and cause systemic effects. Studies indicate particles smaller than 1 μm in diameter may penetrate intact skin (Tinkle et al. 2003; Ryman-Rasmussen et al. 2006). Dermal irritation has been seen following topical application of single-walled carbon nanotubes (SWCNTs) to nude mice (Shvedova et al. 2003; Murray et al. 2007), although it is not known whether skin penetration could occur and result in adverse health effects. Additional data are needed to extrapolate these findings for identifying any occupational health risks and for investigating the dermal toxicity of other nanomaterials.

Ingestion of nanomaterials might occur due to unintentional hand-to-mouth contact, thereby allowing possible transfer to other body organs via the gastrointestinal tract. The mucociliary escalator system, where particles that are deposited in the lung are transferred by coughing to the pharynx and subsequently swallowed, is an additional path to ingestion. Little is known about possible adverse effects from ingestion of nano-materials; however, some evidence indicates smaller particles can be transferred across the intestinal wall more readily than larger particles (Behrens et al. 2002).

5. Exposure Assessment

An exposure assessment should identify tasks that may expose workers to nanomaterials and also identify the researchers conducting those tasks. Such an assessment would review the process and material flow plans for the facility and the status of specific projects. It would include staff interviews and a walk-through of the facility (laboratory) to ensure that all activities and potential exposure pathways are identified. The inventory of tasks and workers should include information on the potential magnitude, duration, and frequency of exposure during different job tasks, or at specific processes, and the amounts of materials being used. Current work practices and existing engineering controls should be evaluated.

The work tasks should be inventoried and prioritized according to the potential for occupational exposure. Examples of tasks and product activities include the following:

- Material receipt, unpacking, and delivery.
- Laboratory operations (synthesis, analytical, and quality assurance activities).
- Cleaning and maintenance.
- Storage, packaging, and shipping.
- Reasonably foreseeable emergencies.
- Waste management.

Determinants of potential exposure to nanomaterials may include dustiness, type of process, quantity of material handled, and duration and frequency of employee exposure. These elements are summarized below and should be taken into account when implementing exposure control measures.

Dustiness

The dustiness of the nanomaterial can influence potential exposures and the selection of the appropriate engineering control. Dustiness describes the tendency of particles to become and/or stay airborne and refers not only to the physical form of the nanomaterial but also to the electrostatic repulsive forces inherent in the particle. For example, the "dustiness" of the nanomaterial is influenced by its particle bulk density and morphology (shape, diameter, and length), as well as the incorporation of the nanomaterial into slurries or liquid suspensions. Nanomaterials in dry powder form tend to pose the greatest risk for inhalation exposure, while nanomaterials suspended in a liquid typically present less risk via inhalation.

Exceptions have been identified during some laboratory processes such as sonication, which resulted in an increase in airborne nano-materials (Johnson et al. 2010). Electrostatic forces influence the stability of particle dispersion in air. These electrostatic forces therefore affect dustiness and should be controlled where possible. Nanomaterials with little or no repulsive forces will tend to be more likely to form aggregates and therefore be less dusty. Nanomaterials incorporated into a solid matrix present the least risk for inhalation exposure because of their limited mobility as long as they are maintained within the matrix.

Process

Some material handling, synthesis, and manufacturing processes can increase the risk of employee exposure. Open, manual handling of bulk nanomaterials, as well as high-energy processes such as milling, sonication, grinding, and high-speed blending, could cause the release of nanomaterials (Gohler et al. 2010; Johnson et al. 2010). Consideration should also be given to the possibility of intentional or inadvertent chemical changes during a work task that may alter the toxicity of a nanomaterial.

Quantity, Duration, and Frequency of Task

The quantity of the nanomaterial that is synthesized, received, or handled in the laboratory will significantly influence exposure potential. Research laboratories may handle quantities ranging from milligrams to several grams or even kilograms of a nanomaterial. As quantities increase, consideration of additional control measures may be required. Exposure potential may be influenced by the duration and frequency of the task(s). Small quantities used on an infrequent basis may not require the same level of control measure that large quantities used daily would require.

Engineering controls should be the primary means of controlling exposures, except in situations (e.g., emergencies) where such controls may not be feasible. In those circumstances, other control measures may be required (e.g., respirator use).

5.1. Safety through the Life Cycle of a Nanomaterial

To ensure the health and safety of those working with nanomaterials, the exposure sources during the nanomaterial product life cycle should be evaluated. Exposure sources include nanomaterial synthesis reactors, nanoparticle collection and handling, product fabrication with nanomaterials, product use, and product disposal (Sahu and Biswas, 2010). Table 1 contains some selected activities with potential exposure sources and recommended engineering controls. The ultimate disposal of the nanomaterial and contaminated refuse should follow all applicable federal, state, and local regulations.

Consideration should be given to installing high-efficiency particulate air (HEPA) filters on laboratory chemical hoods or other individual exhaust duct systems. The decision to use HEPA filtration should be based on evaluation of the contaminant characteristics, maintenance and protection of the fan motor and other exhaust parts, energy requirements, and any applicable environmental release rules. NIOSH recommends the use of HEPA filtration on local exhaust ventilation, laboratory chemical hoods, low-flow enclosures, and any other containment enclosures as a best practice during the handling of engineered nanomaterials.

5.1.1. Synthesis

Nanomaterial synthesis represents the first step in the nanomaterials exposure pathway. Numerous methods can be used to synthesize nanomaterials, and the nature of the potential exposure depends on the specific synthesis process and the stage within the process. Appropriate engineering controls will depend greatly on the synthesis method utilized and the step in the process where the exposure might occur.

Table 1. Employee activities and recommended minimum controls

State of the nanomaterial	Employee activity	Potential exposure source	Recommended engineering controls
Bound or fixed nanostructures (polymer matrix)	■ Mechanical grinding, alloying, etching, lithography, erosion, mechanical abrasion, grinding, sanding, drilling, heating, cooling	■ Nanomaterials may be released during grinding, drilling, and sanding. Heating or cooling may damage the matrix, allowing release of nanomaterial.	Local exhaust ventilation ■ Laboratory chemical hood (with HEPA-filtered exhaust) ■ HEPA-filtered exhausted enclosure (glovebox) ■ Biological safety cabinet class II type A1, A2, vented via thimble connection, or B1 or B2
Liquid suspension, liquid dispersion	■ Synthesis methods: chemical precipitation, chemical deposition, colloidal, electrodeposition crystallization, laser ablation (in liquid) ■ Pouring and mixing of liquid containing nanomaterials ■ Sonication ■ Spraying ■ Spray drying	■ Exposures may result from aerosolization of nanoparticles during sonication or spraying, equipment cleaning and maintenance, spills, or product recovery (dry powders).	■ Laboratory chemical hood (with HEPA-filtered exhaust) ■ HEPA-filtered exhausted enclosure (glovebox) ■ Biological safety cabinet class II type A1, A2, vented via thimble connection, or B1 or B2
Dry dispersible nanomaterials and agglomerates	■ Collection of material (after synthesis), material transfers, weighing of dry powders, mixing of dry powders	■ Exposures may occur during any dry powder handling activity or product recovery.	■ Laboratory chemical hood with HEPA-filtered exhaust ■ HEPA-filtered exhausted enclosure (glovebox) ■ Biological safety cabinet class II, B1 or B2

State of the nanomaterial	Employee activity	Potential exposure source	Recommended engineering controls
Nanoaerosols and gas phase synthesis (on substrate)	■ Vapor deposition, vapor condensation, rapid solidification, aerosol techniques, gas phase agglomeration, inert gas condensation (flame pyrolysis, high temperature evaporation), or spraying	■ Exposures may occur with direct leakage from the reactor, product recovery, processing and packaging of dry powder, equipment cleaning, and maintenance.	■ Glovebox or other sealed enclosure with HEPA-filtered exhaust ■ Appropriate equipment for monitoring toxic gas (e.g., CO)

Table adapted from the summary of recommended nanomaterial controls from the University of New Hampshire (UNH 2009), the University of North Carolina (UNC 2011), and the Research Report 274 (Aiken et al. 2004) prepared by the Institute of Occupational Medicine for the Health and Safety Executive.

Because of the possibility of equipment leaks, synthesis processes should be carried out in an isolated area or in an enclosure operating under negative pressure and exhausted through HEPA filters (Seaton et al. 2010). Precautions such as local exhaust ventilation and PPE should be utilized when cleaning or performing maintenance on the equipment. Furthermore, general ventilation in the laboratory is often not sufficient to effectively clear nanomaterials released into the general room air over a 30-minute period; therefore, researchers should leave the hood fan on even after synthesis is complete (Sahu and Biswas 2010). Finally, exposures could occur during product recovery, packaging, and shipping phases and therefore should be identified and controlled according to the state of the nanomaterial during that stage of synthesis (Table 1).

Carry out operations in a manner that minimizes the risk of exposure to nanomaterials from inhalation or dermal contact. Principles that contribute to minimizing the risk of exposure to nanomaterials in the laboratory include the following:

- Handle nanomaterials in dry powder form with care to minimize the generation of airborne dust and to minimize dermal contact.
- Nanomaterials suspended in a liquid present less risk for becoming airborne than nanomaterials in dry powder form under normal handling conditions, but they may present a dermal risk, especially if the nanomaterial is suspended in a solvent.
- Nanomaterials suspended in a liquid may be aerosolized during certain handling activities (for example, during sonication).
- Nanomaterials incorporated into a solid matrix are least likely to become airborne because of their limited mobility. However, under certain circumstances these nanomaterials may still pose some risk, such as if the solid matrix is cut, sawed, drilled, sanded, or handled in any way that creates a dust or releases the nanomaterial.
- The quantity of material handled contributes greatly to the risk of exposure. Operations involving the use of nanomaterials should always use the minimum quantity required for the particular experiment or process.

5.1.2. Characterization and Purification

Once the nanomaterial has been synthesized, it may undergo characterization, purification, or other modification steps such as the addition of surface coatings to functionalize it. Safety precautions and standard operating procedures should be developed and followed for hazards associated with the characterization, purification, or functionalization of the nanomaterial.

Characterization includes the determination of the size and shape of the nanomaterial, atomic and electronic structures, and any other important chemical or physical properties (Rao and Biswas 2009). This process may include various analytical methods such as microscopy, X-ray diffraction, and spectroscopy. Purification or processing of nano-materials is used to remove impurities from the nanomaterial of interest. For example, a raw carbon nanotube material may contain the catalyst used in the synthesis process. Purification techniques include high-temperature heat treatments, the application of highly acidic or caustic substances, or the use of potentially hazardous solvents. Functionalization modifies the particle surface by attaching another substance, which may change the toxicity or behavior of the nanomaterial.

5.1.3. Application and Material Testing

The third phase of nanomaterial processing in research laboratories involves the application and testing of the nanomaterial or nanoenabled material. This may involve combining the nanomaterial into other matrices, applying nanomaterials onto surfaces, or destructive testing of substances containing nanomaterials.

Avoid manipulating nanomaterials in open systems or in a free particle state (e.g., handling dry nanopowders on a bench top).

- Preferably, (1) keep nanomaterials bound in a matrix, (2) keep them suspended in a liquid, (3) keep them sealed in a container, or (4) use appropriate engineering controls.
- For larger processes that cannot fit in a fume hood or glovebox (e.g., injection molding), control emissions with properly designed local exhaust ventilation.
- Transfer nanomaterial samples between workstations (such as exhaust hoods, gloveboxes, furnaces) in sealed, unbreakable, labeled containers.
- Avoid generating nanoparticle aerosols (e.g., through sonication) on bench tops. Use appropriate laboratory exhaust and containment systems.

The most stable form for most nanomaterials occurs when they are bound within a solid matrix. However, destructive treatment of the matrix, such as during grinding, sanding, and drilling, may lead to the release of nanoparticles or larger particles containing nanomaterial. Thermal stresses such as melting plastics may also cause nanomaterial release. Local exhaust ventilation

(Section 7.0) should be used during destructive handling of matrices containing nanomaterials (UNH 2009).

Working with nanomaterials in liquid during activities where energy is applied such as sonication or mixing may generate the airborne release of respirable droplets containing nanomaterials (Johnson et al. 2010). Proper controls should be used during these operations (Table 1).

6. Recommendations for Exposure Control

Among the most effective means to prevent occupational injuries and illnesses are anticipating potential occupational safety and health hazards early in the development of the technology or process and incorporating safe practices into all design, implementation, and operation phases. Prevention through Design (PtD) is a management tool for protecting workers from potentially unsafe work conditions. It emphasizes the importance of employee health and safety through the design, construction, manufacture, use, maintenance, and ultimate disposal or reuse of tools, equipment, machinery, substances, work processes, and work premises (NIOSH 2010b). PtD addresses occupational safety and health needs by eliminating hazards and minimizing risks to workers throughout the life cycle of the process (Figure 2) (Schulte et al. 2008b). Many nano-technology research laboratories recognize PtD as a cost-effective means to enhance occupational safety and health and have incorporated PtD management practices within their facilities (Murashov and Howard 2009).

Prevention through Design strategies follow the standard hierarchy of controlling workplace hazards, which includes (1) eliminating, substituting, or modifying the nanomaterials; (2) engineering the process to minimize or eliminate exposure to the nanomaterials; (3) implementing administrative controls that limit the quantity or duration of exposure to the nanomaterials; and (4) providing for use of PPE.

6.1. Elimination or Substitution

For nanomaterial researchers, it is often not feasible to eliminate or substitute the nano-material. It may be possible, however, to change some aspects of the process in a way that reduces release of the nanomaterial. For example, working with nanomaterials suspended in a liquid is a significant improvement over working with them in dry powder form, because the potential for airborne release is reduced in most laboratory processes.

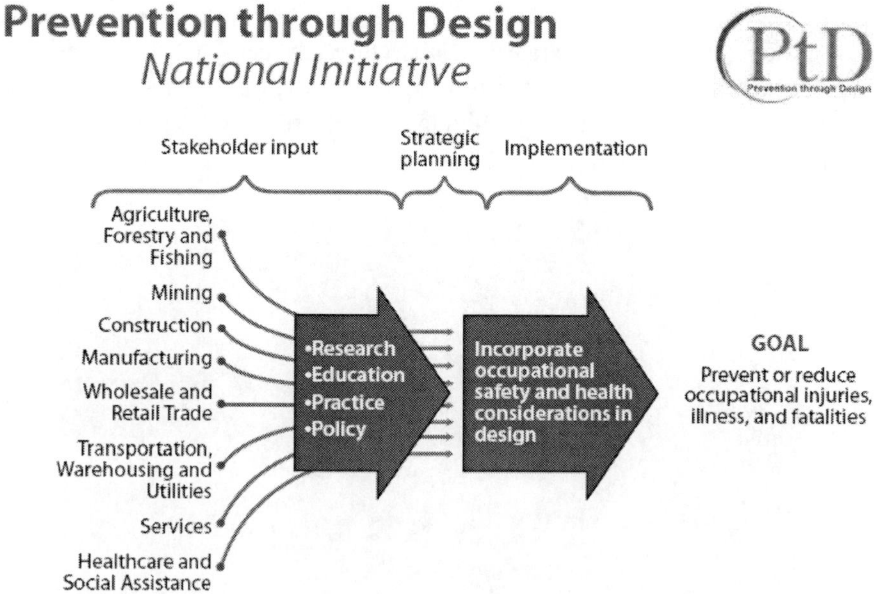

Figure 2. Prevention through Design.

However, physical agitation of the liquid (e.g., sonication) may aerosolize small droplets containing the nanomaterial (Johnson et al. 2010) (Figure 3).

Opportunities for eliminating the use of hazardous materials or substituting for less hazardous forms do exist in other aspects of nanomaterials production. Engineered nanoparticle research often requires the use of solvents and other potentially hazardous chemicals.

A recent article on optimizing the properties of carbon nanotubes (CNTs) reported that eight different solvents were evaluated for the optimization; all eight solvents, including the solvents benzene, toluene, ethyl acetate, and dimethylformamide, were considered toxic to different degrees (Ju et al. 2009). Researchers should always attempt to identify and use chemical processes that utilize nontoxic or less-toxic alternatives whenever possible, in order to minimize worker exposures and environmental releases when the process is scaled up to full production. This control strategy, substituting a less toxic material in production processes, has been the focus of much research during the past 20 years. One source for valuable information on process change or chemical substitution is the Toxics Use Reduction Institute at the University of Massachusetts Lowell (www.turi.org).

It is also possible to substitute a less "energetic" operating condition, and thereby modify a process to make it inherently safer. An example of process modification was demonstrated in a laboratory producing CNTs by chemical vapor deposition. Optimizing the furnace reaction temperature maximized the production of CNTs while minimizing the release of CNTs in the furnace exhaust (Tsai et al. 2009b).

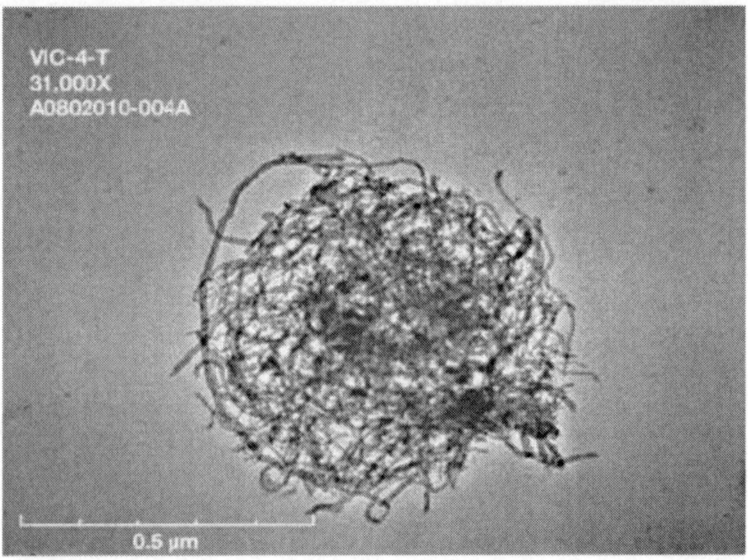

Figure 3. Aerosol droplets containing nanomaterials ejected from vial during sonication.

6.2. Isolation and Engineering Controls

Isolation includes the physical isolation of a process or piece of equipment either by locating it in an area separate from the worker or by placing it within an enclosure that will contain the nanomaterials released. Engineering controls include any physical change to the process or workplace that reduces contaminant emissions and subsequent employee exposure. Several factors will influence the selection of exposure controls for nanomaterials, including quantity of nanomaterial handled or produced, physical form, and task duration. As each one of these variables increases, exposure risk becomes greater, as does the need for more efficient exposure control measures (Figure 4, adopted from NIOSH (2009a)). Operations involving easily dispersed dry nanomaterials deserve more attention and more stringent controls (e.g., enclosure) than those where the nanomaterials are suspended in a liquid matrix or imbedded in a solid. Liquid nanoparticle suspensions rarely pose a danger of inhalation exposure during routine operations, but they may represent a significant hazard when aerosolized or in unexpected situations such as a spill. Nanomaterials incorporated into bulk solids may pose some risk if the solid matrix is cut, sawed, drilled, sanded, or handled in any way that creates a dust or releases the nanomaterial.

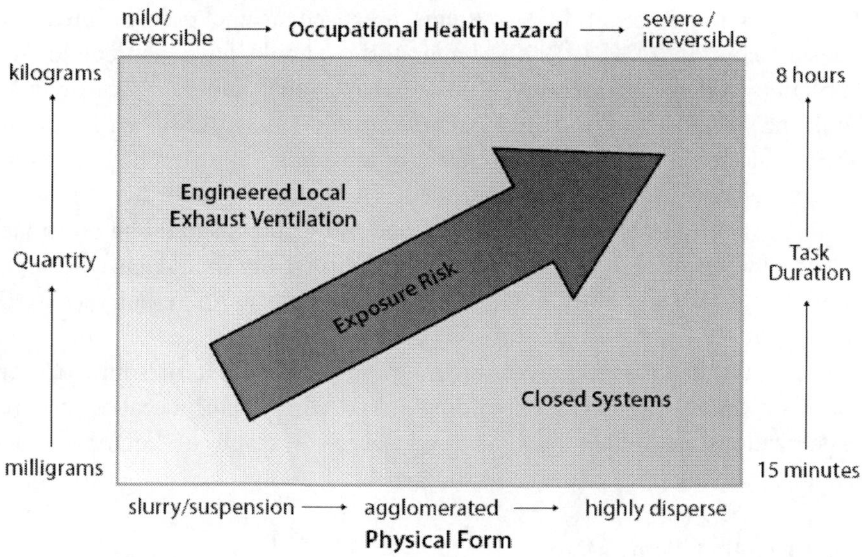

Figure 4. Factors influencing control selection.

6.2.1. Containment

Containment refers to the physical isolation of a process or a piece of equipment to prevent the release of the hazardous material into the workplace. An example of process isolation would be the location of a twin-screw extruder used to make CNT composites in a room separated from the rest of the research facility. An example in chemistry labs is the use of specially designed separate storage cabinets for flammables, acids, and bases. Another example of containment would be a glovebox, which is a sealed container with attached gloves that allows the researcher to carry out process or tasks while being physically separated from the hazard.

6.2.2. Ventilation

General Exhaust Ventilation (GEV)

It is important that any laboratory working with nanomaterials have sufficient general exhaust ventilation (GEV); however, it should not be the sole means of controlling nano-material exposure. GEV is typically provided by the building's heating, ventilation, and air conditioning (HVAC) system. Recommended ventilation rates for general laboratory use range from 4 to 12 air changes per hour, if LEV systems are used as the primary means of exposure control (OSHA 1990). Laboratories should have nonrecirculating ventilation systems (preferably, 100% exhaust air), and lab pressurization should be negative to the hallway (DiBerardinis 1993). Additionally, the air supply and air exhaust should be carefully located so that supplied air passes through the area that is being controlled. The exhaust should be as close as possible to the source of contamination, and the workers should be positioned between the air supply and the source. Exhausted air should be discharged away from windows, other air intakes, or other means of re-entry (ACGIH 2007).

Care must be taken to prevent the migration of nanomaterials into adjacent rooms or areas through the building's HVAC system, because of area pressurizations and directional airflows, or as a result of equipment and personnel moving from one area to another.

Local Exhaust Ventilation

A local exhaust ventilation (LEV) system with air cleaner is shown in Figure 5. Laboratory settings would have chemical fume hoods, vented enclosures, and special devices connected. The exhaust hood typically is next to or encloses the contaminant source to control exposures at the source. Air flowing into the hood entrains the contaminants and carries them through the duct, where they are either removed by an air cleaner or vented to the atmosphere. Other LEV systems include biological safety cabinets and powder-handling enclosures. Section 7.0 is dedicated entirely to LEV and its use with nanomaterials.

Use good housekeeping in laboratories where nanomaterials are handled.

- Clean all working surfaces potentially contaminated with nanomaterials (e.g., benches, glassware, apparatus, exhaust hoods, support equipment) at the end of each day with a HEPA vacuum and/or wet wiping. Do not dry sweep or use compressed air.
- As an alternative to HEPA-vacuuming lab bench tops, bench top protective covering material may be used.
- Make use of hand-washing facilities before eating, smoking, or leaving the worksite.
- Use facilities for showering and change clothes to prevent the inadvertent cross-contamination of other areas (including take-home).
- Provide laundry service for contaminated work clothing.
- Do not eat or drink in the areas where nanomaterials are handled.
- Collect laboratory waste in sealed, labeled containers approved for the particular waste stream in a manner that minimizes potential exposure during the transfer of waste into the container. Store the container in secondary containment.

6.3. Administrative Controls

Administrative controls contribute to worker exposure reduction, but they do not always reduce the airborne concentration of the contaminant in the workplace. They often include limiting exposure by reducing the time the employee is handling the material, specifying good housekeeping and other good work practices, training employees, and implementing proper labeling and storage of materials. Administrative controls in some research laboratories may include maintaining clean room conditions (Schulte et al. 2008).

Hand-washing facilities should always be used before eating, drinking, smoking, or leaving the workplace. Food and drink should not be permitted in the areas where nanomaterials are handled.

6.3.1. Employee Training

All employees working with engineered nanomaterials should receive training on the associated hazards and risks. The OSHA Hazard Communication Standard, 29 CFR 1910.1200, requires that at a minimum, training should address means to detect the chemicals in the workplace, the hazards associated with those chemicals, and procedures to prevent exposure (OSHA 1994). In addition training should include appropriate nanomaterial handling and storage procedures, proper use of PPE, cleaning of contaminated surfaces or clothing, and proper disposal of nanomaterials or nanomaterial-contaminated objects (NIOSH 2008). Employees should be educated regarding the job tasks that may expose them to nanomaterials and the use of appropriate controls and work practices to minimize exposure.

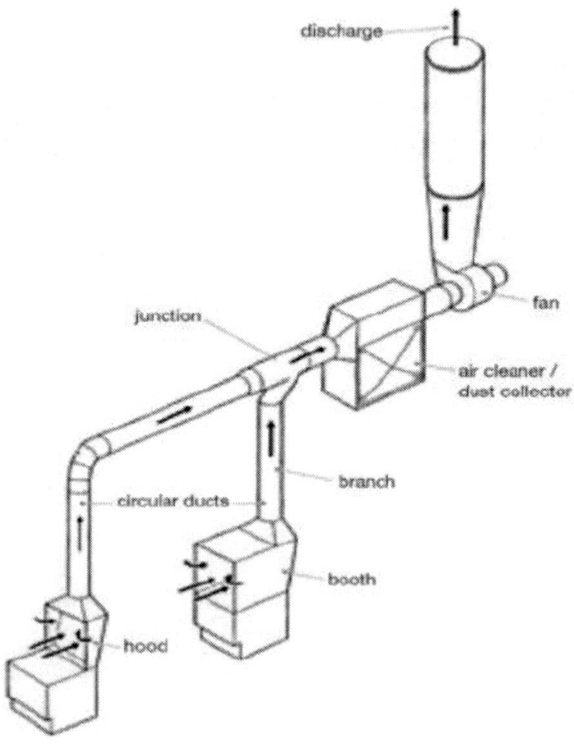

Figure 5. A local exhaust ventilation system with air cleaner.

When working with nanomaterials, use space that is isolated as much as possible from the rest of the lab, with as few people in that space as possible.

- Keep laboratory doors closed and limit access to the laboratory (e.g., via key cards) to prevent unauthorized access.
- Post appropriate warnings in laboratories, including measures to be taken to protect laboratory researchers and visitors from exposure risks.
- Use local exhaust in all areas of material collection and transfer where possible.
- Cover all containers when not in use.
- Use 100% fresh supply air. Do not recirculate room air.

6.3.2. Labeling and Storage

Under the OSHA Hazard Communication Standard, 29 CFR 1910.1200, employers are required to label all hazardous chemicals in the workplace. Nanomaterials should be stored in labeled containers that indicate their chemical content and form. Liquids or dry particles should always be stored in unbreakable, tightly sealed containers. Secondary containment should be used when appropriate. Appropriate signage indicating the hazard, PPE requirements, and any other pertinent information should be posted at entry points to areas where nanomaterials and other hazardous compounds are handled or stored.

6.4. Personal Protective Equipment

Personal protective equipment (PPE) should be required when engineering and/or administrative controls are not feasible or effective in reducing exposures to acceptable levels and wherever it is necessary because of hazards. Protective equipment must be used and maintained in a sanitary and reliable condition (OSHA 2008). Based on the uncertainty of the health risk of nanomaterials, it may be prudent to wear appropriate PPE on a precautionary basis. PPE can include respirators, gloves, clothing, face shields, safety glasses, and other garments designed to protect the wearer.

6.4.1. Protective Clothing

There are no standards or guidelines for the use or selection of protective clothing or other apparel for working with nanomaterials (OSHA 2008). Suggested PPE consistent with basic industrial hygiene practice includes the following:

- Clothing appropriate for a wet-chemistry laboratory, including closed-toe shoes made of a low permeability material. (Disposable, over-the-shoe booties may be necessary to prevent tracking nanomaterials from the laboratory).
- Long pants (without cuffs) and a long-sleeved shirt.
- Impervious laboratory coats (noncotton). (If nondisposable laboratory coats are used, they should remain in the laboratory/change-out area to prevent nanoparticles from being transported into common areas).

- All re-useable protective clothing should be laundered. The clothing should be placed in closed bags before being taken out of the laboratory for cleaning in a central, approved location.
- Safety glasses/goggles and/or face shields as appropriate, as determined in an assessment of the hazard risk. A face shield alone is not sufficient protection against unbound dry materials.
- Nitrile or other chemically impervious gloves, as appropriate for handling nano-material powders and liquids. Suggested guidelines for the selection and use of gloves are as follows:
    - The proper selection of gloves should take into account the resistance of the glove to the nanomaterial (if available) and, if the nanomaterial is suspended in liquids, the liquid.
    - Chemically resistant gloves can develop cracks when they are used, so gloves should be changed whenever they show visible signs of wear.
    - Contaminated gloves should be kept in a closed plastic bag in the work area until disposal.
    - If protective clothing and/or gloves are required, particular attention should be given to preventing exposure to skin, especially abraded or lacerated skin. (Figure 6 demonstrates improper and proper use of sleeves).
    - Special attention should be given to the proper removal and disposal of contaminated gloves to prevent skin contamination.
    - Gloves should also be routinely replaced to minimize the risk of exposure and contamination of the work environment.

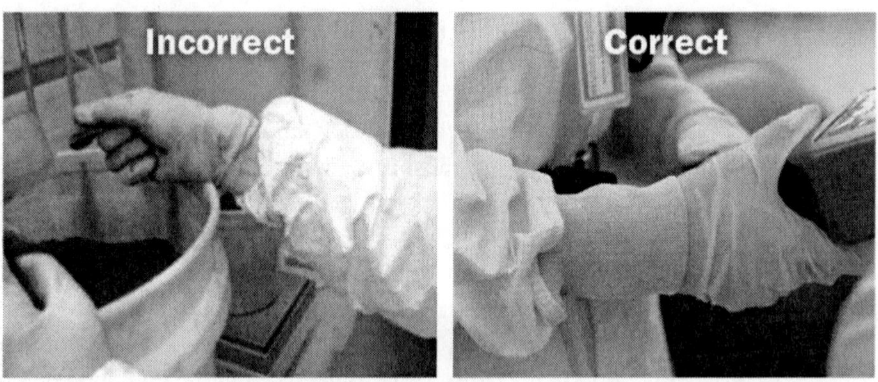

Figure 6. Make sure that the coveralls and gloves do not allow nanomaterials to contact the skin.

6.4.2. Respirators

When the potential exists that workers may inhale nanomaterials due to a lack of effective engineering controls or during activities with higher nanomaterials exposure potential (e.g., emergencies), appropriate respirators, selected according to the NIOSH Respirator Selection Logic (NIOSH 2005), should be used pursuant to the Occupational Safety and Health Administration (OSHA) respiratory protection standard 29 CFR 1910.134 (OSHA 1992). Figure 7 shows an example of a process where a respirator was worn when working with nanomaterials.

The OSHA respiratory protection standard requires the development of a written respiratory protection program for any workplace where respirators are necessary to protect the health of the worker or whenever required by the employer. The program should include the following elements (OSHA 1992):

- Procedures for selecting respirators for use in the workplace.
- Medical evaluations and fit-testing of employees required to use respirators.
- Procedures and schedules for cleaning, disinfecting, storing, inspecting, repairing, discarding, and otherwise maintaining the respirator.
- Training of employees in the respiratory hazards to which they may be exposed, and proper use and maintenance of the respirator.
- Procedures to evaluate the effectiveness of the program.

The NIOSH proposed recommended exposure limit (REL) of 7 $\mu g/m^3$ elemental carbon as an 8-hr Time Weighted Average respirable mass airborne concentration for carbon nano-tubes and carbon nanofibers, and 0.3 mg/m^3 for ultrafine titanium dioxide suggest respirators may be necessary if expected exposures are above this level (NIOSH 2010a, 1994, 2011). A properly fit-tested, half-face particulate respirator will provide protection at exposure concentrations 10 times the REL, while an elastomeric full face piece respirator with P100 filters will provide protection at 50 times the REL. NIOSH provides further guidance for selecting respirators in the NIOSH Respirator Selection Logic 2004 (NIOSH 2005).

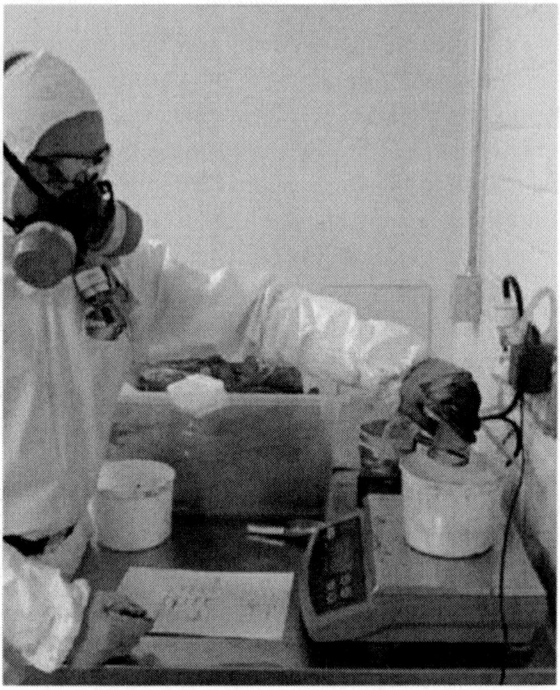

Figure 7. Weighing of carbon nanotubes.

Personal protective equipment should be used when there is the potential for exposure to the nanomaterial.

- At a minimum, for all laboratory activities, wear impervious (non-woven) laboratory coats (or coveralls, or a work uniform that covers the arms); long pants without cuffs: a long-sleeved shirt; closed-toe shoes made of a low-permeability material, or disposable foot covers; eye protection; and appropriate chemical-resistant gloves (depending on the chemical exposure).
- Respiratory and face protection and jumpsuits or chemically resistant protective clothing may be needed for laboratory activities, depending on the hazard or quantities of the material(s) handled, the availability of appropriate controls, and the exposure risks.
- Respiratory protection should be selected in consultation with the laboratory occupational safety and health professional according to the NIOSH Respirator Selection Logic.
- Follow proper procedures to properly select, maintain, don, doff, and decontaminate personal protective clothing and equipment as described in the laboratory risk management plant.
- Provide an area outside of the contaminated area for donning and doffing PPE.

7. Local Exhaust Ventilation

Local Exhaust Ventilation (LEV) systems reduce or prevent exposure to airborne contaminants by capturing them at their source. The first element of a LEV system is the hood, of which there are two basic types: enclosing and exterior (ACGIH 2007). The most common LEV system used in research laboratories is the laboratory chemical hood. An enclosing laboratory chemical hood is shown in Figure 8. The hood, with its moving sash, is actually only a partial, "three-sided" enclosure. Other common types of LEV include the exterior hood (Figure 9), which is placed adjacent to the contaminant source. Due to the unknown hazard potential of nanomaterials, a more conservative approach to ventilation control is dictated, with an emphasis on enclosing systems.

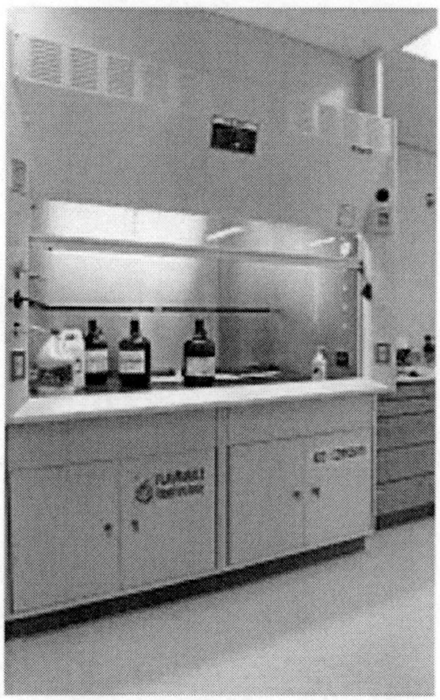

Figure 8. Laboratory chemical hood (note that the work area is *inside* the hood).

In general, enclosing hoods are preferred to exterior hoods, because the contaminants are contained inside the hood itself. The hood provides a barrier between the worker and the contaminant (this barrier is only partial for a laboratory chemical hood when the sash is opened—see the discussion on these hoods below). Sufficient airflow must be provided through any openings in the enclosure to ensure that the contaminants don't escape the hood. Airflow through openings is usually specified as a certain required face velocity at the opening, and the value depends on the hood design and application (ACGIH 2007). It can be difficult to choose the proper face velocity to achieve adequate flow for enclosing hoods with large openings. Most laboratory chemical hoods are designed to operate at 100 ft/min face velocity, which can create problems because of turbulence. Turbulence may reduce the capture efficiency of the hood and may disturb settled particles (NRC 2011). This is discussed more fully in the following section. Exterior hoods are less preferred because they must create a capture velocity at the point of contaminant generation to capture the contaminant and draw it into the hood.

Figure 9. Exterior hood (note that the work area is *in front* of the hood entrance).

7.1. Laboratory Chemical Hoods

A properly designed and maintained chemical fume hood can offer significant worker protection if used properly. There are many different hood designs, but the most common categories are the conventional or constant-flow hood, the bypass hood, and the variable air volume constant- velocity hood. Examples of each are shown in Figures 10–12. Positive pressure laminar flow hoods that are designed for product protection and direct potentially contaminated air toward the user are not considered local exhaust ventilation and are not recommended for use to protect from nanomaterial exposures (Harford 2007). All chemical hoods have certain common design elements, including an exhaust fan to move air through the hood, a moving sash, exhaust slots, and a horizontal work surface. The sash can move in either a vertical or a horizontal direction. A crucial performance element for any chemical hood is the face velocity, defined as the average air velocity at the face of the hood at the sash opening. Maintaining a constant, minimum face velocity provides confidence that operations (and hazardous agents) within the hood will be contained. Hood face velocity must be evaluated and controlled by the facility's engineering or health and safety staff. The current consensus of the literature is that the average face velocity for a laboratory chemical hood should be in the range of 80–120 ft/min (Burgess et al. 2004). The flow control system on a constant-velocity hood should be adjusted to give a face velocity in this range. Each chemical hood should be clearly marked with the proper hood sash location that will give the desired face velocity; depending on the hood design, this could be a single location or a range of locations. Containment verification using tracer gases to provide quantitative data and smoke testing to visual airflow patterns is recommended when the hood is installed, when substantial changes are made to the ventilation system, and periodically as part of a preventive maintenance program. Testing should be performed following the ANSI/ ASHRAE 110 or equivalent protocol (NRC 2011).

In addition to the face velocity, it is important that the airflow be distributed evenly across the hood face. ANSI/AIHA Z9.5 recommends that variations of face velocity across the hood face should be within ±20% of the average face velocity; however, some laboratories select a stricter standard of ±10%.

The constant-flow hood (Figure 10) constitutes the oldest, simplest chemical hood design. The exhaust fan introduces a constant volumetric airflow moving through the sash opening. For this hood design, the face velocity is lowest when the sash is wide open; when the sash is lowered the face velocity increases.

The bypass hood (Figure 11) maintains a constant hood face velocity and incorporates a bypass grille located above the sash opening. When the sash is wide open it blocks the bypass grille, allowing all of the air to flow through the hood opening. As the sash is lowered, it uncovers increasingly greater amounts of the bypass grille, allowing increasing amounts of air to flow through this alternative path. If it is designed and operated properly, the amount of air flowing through the bypass grille is just sufficient to maintain a constant face velocity. Typically, however, this constant velocity can be maintained over a certain part of the sash's total range.

The constant-velocity hood (Figure 12) uses a control system to detect the sash position, face velocity and system pressure, and change the fan motor speed or other mechanism, such as mechanical dampers, to increase the airflow when the sash is raised and decrease it when the sash is lowered, thus maintaining a constant face velocity.

Tsai et al. (2010) evaluated the efficiency of these three main types of laboratory chemical hoods to reduce exposure to aluminum oxide nanomaterials while manually handling them inside the chemical hoods. They determined that the particle release to the worker's breathing zone was greatest when using a constant-flow hood, as compared to a bypass and a constant-velocity hood.

Figure 10. Conventional (constant-flow) laboratory chemical hood.

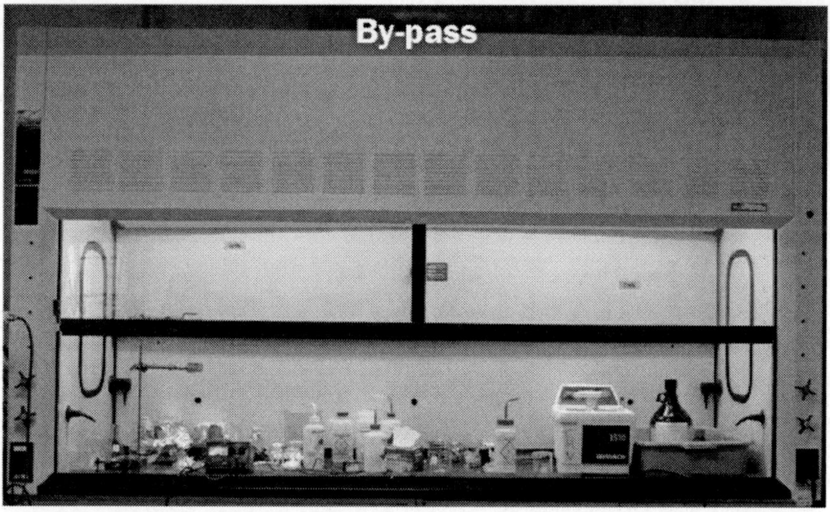

Figure 11. Laboratory bypass hood (note the by-pass chamber above the sash).

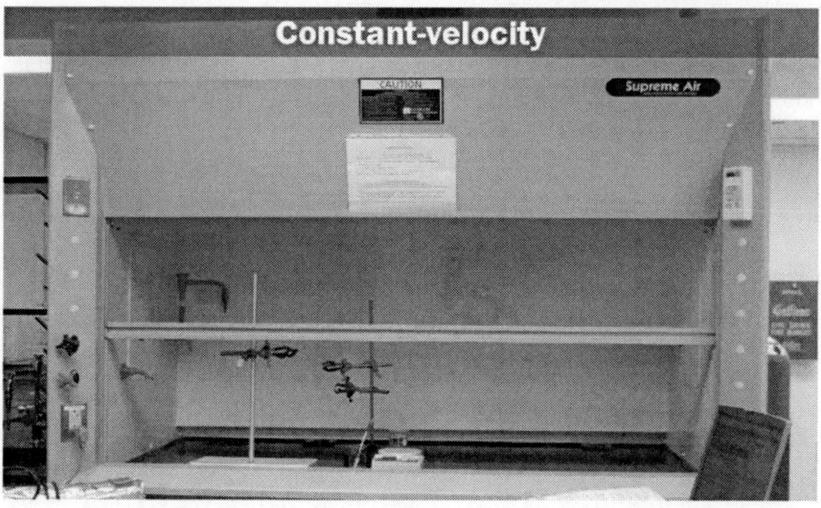

Figure 12. Laboratory constant-velocity hood (note the flow monitor on the sash column).

Saunders (1993) and DiBerardinis (1993) described basic elements that all chemical hoods should incorporate. All laboratory chemical hoods should include, *at a minimum*, the following design elements:

- A minimum width of 4 feet (wider is better, to allow flexibility in equipment use).
- A minimum sash open height of 30 inches.
- A bottom-front airfoil.
- A sash that is easily movable over its entire range of motion.
- A sash that holds its position over its entire range of motion.
- Side walls that are smooth, rounded, and tapered toward the inside of the sash opening (Schulte et al. 1954).

In addition, the following factors relative to the hood location are very important for proper hood performance

- Air currents outside a hood may disrupt the airflow at the face and therefore impact the ability of a hood to contain the contaminant.
- The hood should not be located next to any laboratory entry door or any other high-traffic location.
- The hood should be at least 5 feet from any HVAC air supply grille; a distance of 10 feet is preferred.

The following practices are important for working in laboratory chemical hoods:

- The hood sash should be kept wide open during equipment set-up only; during actual use, the sash should be lowered to the position that gives proper hood face velocity.
- Equipment should be at least 6 inches behind the sash opening (many hoods have a recessed floor starting at this distance, to encourage proper use).
- When working in the hood, the user should avoid working at the edge of the hood and should minimize arm movements; all such movements should be slow and smooth.
- Traffic past the hood should be minimized when nanomaterial powders are being manipulated. Research has shown that the passage of a person past the hood face at walking speeds creates a turbulent wake sufficient to pull contaminants from the hood (Johnson and Fletcher 1996).
- During experiments, when no access is required, the sash should be kept either in the same position as when work is performed (constant

flow and bypass hoods) or lowered to the fully closed position (constant velocity hoods).

- When using a local exhaust system, do not directly exhaust into the work environment any effluent (air) that is reasonably suspected to contain nanomaterials. The exhaust air should be passed through a HEPA filter (NIOSH 2007) and, when feasible, released outside the facility. If the exhausted air is recirculated, then steps should be taken to ensure that recirculated air doesn't contain the engineered nanoparticle.
- Handle exhaust filters from the chemical hoods in a manner that minimizes exposure. Put a plastic-lined bag around the filter at the source when removing it so that particulates are not potentially released to the work environment. Wear appropriate PPE during all maintenance and cleaning activities.
- Storage of materials in the chemical hood should be minimized or eliminated. Materials stored in the hood can adversely affect the containment by disrupting airflow. If items must be placed inside the hood, make sure they are placed near the back and do not block the air slots.

7.1.1. Working with Nanomaterial Powders in Chemical Hoods

Research performed at the University of Massachusetts Lowell (Tsai et al. 2009a) has demonstrated that nanomaterial powders may be released back into the work area from chemical fume hoods during tasks such as weighing or transferring from container to container. Releases that are not detectable on a mass basis were found to have a very high particle number concentration. Experiments performed on constant-volume and bypass hoods demonstrated that working with the sash either too low or too high could cause nanoparticles to escape from the hood. When the sash is too high, the face velocity can fall below the recommended minimum of 80 ft/min. This low face velocity and the large opening created by the high sash allow random room air currents to enter the hood, entrain airborne nanomaterials, and carry them out of the hood. When the sash is too low, the face velocity can exceed the recommended maximum of 120 ft/min. This causes a strong turbulent wake in the space between the worker and the hood face, which can pull airborne nanomaterials from the hood. Because of the possibility of loss of the nanomaterial at high face velocities, the correct sash height should be determined for the specific process being carried out, based on the ability of the chemical hood to capture the nanomaterial. Because of the potential to create turbulence, the hood should be as uncluttered as possible, and the researcher should remove his arms or other objects from the hood very slowly (Tsai et al. 2009a). If the potential for material loss exists or if exhaust filtration is infeasible, alternative exhausted enclosures should be considered such as low flow enclosures or biological safety cabinets (see section 7.2).

7.1.2. New Hood Designs

Researchers are designing new lower-flow chemical hoods that may offer improved performance for handling nanoparticle powders. In some studies, it has been noted that hood face velocities of 100 ft/min may result in the loss of nanomaterial (Johnson et al. 2010). Lower-flow chemical hoods operate with face velocities of less than 100 ft/ min. However, at this time, there is very little research on the effectiveness of low-flow fume hoods for handling nanoparticle powders. A recent hood design approach is the air-curtain hood (Huang et al. 2007), which uses a downward air jet emanating from a double-pane sash to isolate the interior of the hood from the exterior environment. An evaluation of the hood at the University of Massachusetts Lowell (Tsai et al. 2010) indicated that it can be effective at containing airborne nanoparticles.

7.2. Alternatives to Conventional Chemical Hoods

7.2.1. Glovebox Enclosures

A higher level of protection for handling dry powders is obtained by using a glovebox enclosure (DiBerardinis 1993) (Figure 13). The primary advantage of using a glovebox is the protection it affords; when used properly to manipulate nanoparticle powders, a glovebox should prevent exposure to the user. The disadvantages of using a glovebox relate to the extra time required to move materials and equipment in and out of the enclosure, the difficulty of manipulating nanomaterials when wearing gloves, and the need to periodically clean the enclosure. The two most likely sources of exposure when using a glovebox are the transfer of materials into and out of the box and the cleaning of the box following its use. Both of these activities must be performed with extreme care. Note that glovebox enclosures are sometimes used under positive pressure with respect to the surrounding room (e.g., as shown in Figure 13), with an inert atmosphere such as nitrogen to reduce the risks of fire, explosion, or oxidation. Such use can increase the possibility of airborne releases from the enclosure. Proper leak-testing procedures, in accordance with the American Glovebox Society Standards, should be followed to verify containment.

7.2.2. Biological Safety Cabinets

Biological safety cabinets (BSCs) serve as a primary means of containment developed for working safely with infectious microorganisms, such as viruses, bacteria, and fungal spores (Chosewood et al. 2009). BSCs are designed to provide personnel, environment, and product protection when appropriate practices and procedures are followed. Nano-materials, whose size range is similar to that of bioaerosols (microorganisms that are suspended in the air), should behave aerodynamically in the same manner. Additionally, the HEPA filtration systems in BSCs should be equally effective in filtering nano-materials and bioaerosols because of primary particle size. Therefore, it is reasonable to assume that these cabinets will offer similar levels of protection against bioaerosols and airborne nanomaterials. Three different classes of biological safety cabinets are defined as follows:

- A Class I biological safety cabinet resembles a chemical hood, with the additional requirement that the exhaust air must be treated before it is discharged to the atmosphere.
- A Class II biological safety cabinet is designed to protect the operator, the product, and the environment (Figure 14). It has an inward airflow through the open sash to protect the operator, a downward flow of HEPA-filtered air to protect the product, and a HEPA-filtered exhaust to protect the environment. Class II cabinets are designed for use against low- to moderate-risk biological agents. The four types of Class II cabinets are defined as A, B1, B2, and B3; each type of Class II cabinet has different air recirculation percentages, and the level of control increases from A to B3. Because air is recirculated in Class II type A and B1 cabinets, tasks involving volatile materials should not be performed in these cabinets.
- A Class III biological safety cabinet is a highly sophisticated glovebox. The sealed enclosure is maintained at a negative static pressure of at least 0.5 inches H_2O, the supply air is HEPA-filtered, and the exhaust air is either double-HEPA-filtered or passed through a single HEPA filter and then incinerated. Class III cabinets are meant for the highest-risk biological agents.

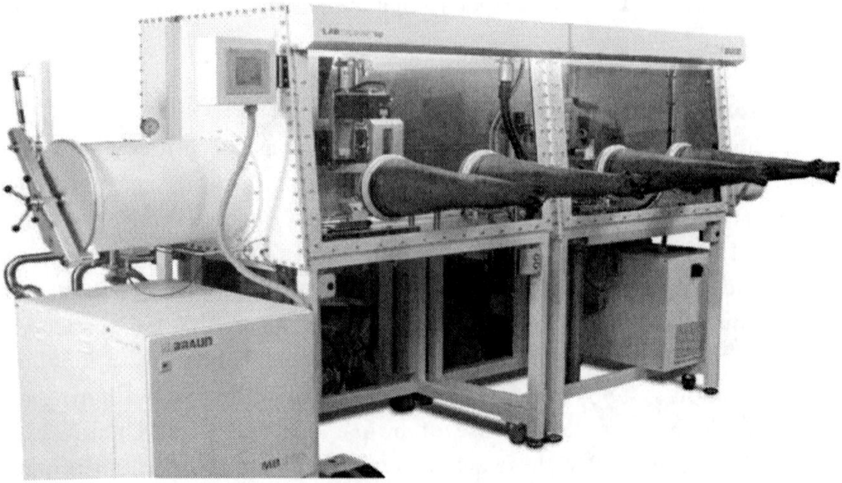

Figure 13. Glove box enclosure (shown here under positive pressure for use with inert atmospheres, rather than for enhanced containment of particles and gases)(MBRAUN; used with permission).

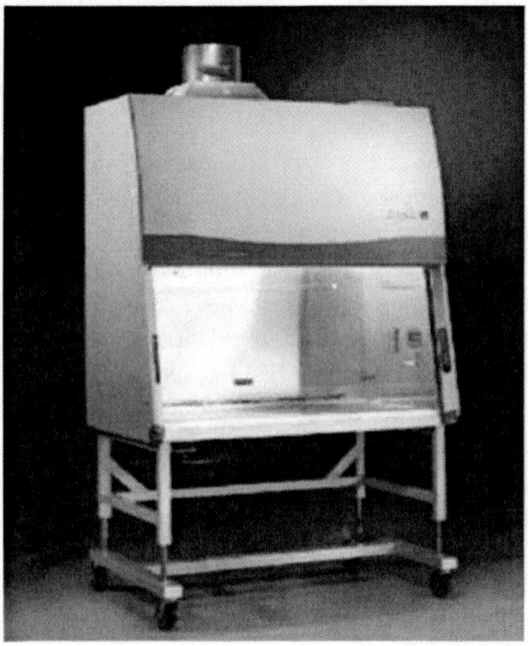

Figure 14. Class II Biological Safety Cabinet (Labconco Corporation; used with permission).

- Duration changes for operations with identified exposure risks.
- A new nanomaterial is handled.
- Physical form changes (for example, powders rather than suspensions).
- New equipment is designed or installed.
- New toxicology data are obtained.
- Medical surveillance trending suggests adverse effects.
- Occupational illness is reported.
- The workforce changes.
- A validated sampling and analytical method is developed for the nanomaterial(s) being used.
- Qualitative indicators (such as odor, visual observations, or employee reports) or quantitative indicators (such as measurements) of exposure suggest a change in control effectiveness.

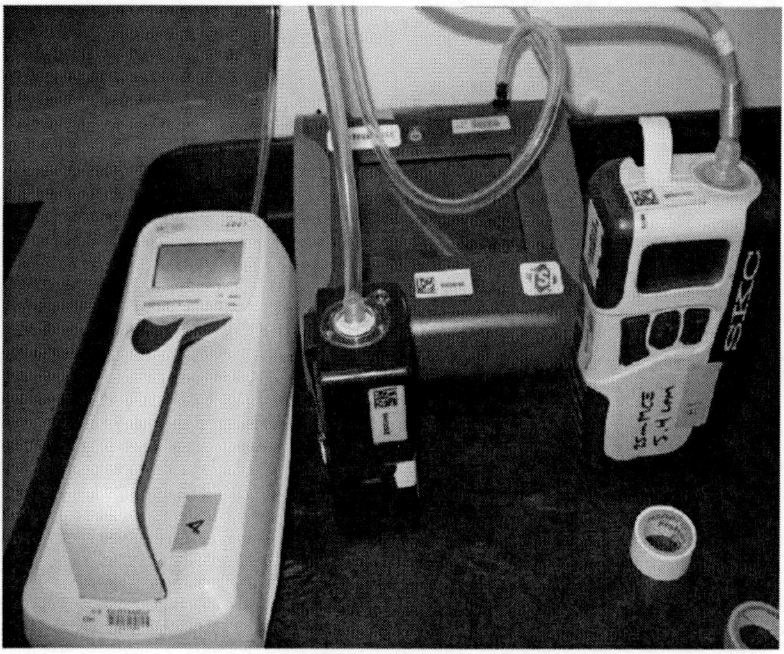

Figure 18. Area sampling for airborne nanomaterials.

10. Guidance on Developing a Control Scheme (Control Banding)

Control banding is a qualitative strategy for assessing and managing hazards associated with chemical exposures in the workplace. The concept is used to manage exposures to potentially hazardous materials through the application of one of four recommended control approaches. This concept is based on the premise that although many chemical hazards exist, there are a limited number of controls available. To determine the appropriate control strategy, one must consider the characteristics of the substance, the potential for exposure, and the hazard associated with the substance. As the potential for harm to the worker increases, so does the degree of control needed to manage the risk (NIOSH 2009c).

The four control bands are usually the following:

Band 1: Use good industrial hygiene practice and general ventilation.
Band 2: Use an engineering control, typically local exhaust ventilation.
Band 3: Enclose the process.
Band 4: Seek expert advice.

There are several control banding tools developed for use with nanomaterial exposures, (Paik et al. 2008; Zalk et al. 2009; *GoodNanoGuide* 2009; Safe Work Australia 2012). The GoodNanoGuide (www.goodnanoguide.org) is an Internet-based platform for the exchange of ideas on handling nanomaterials, and it recommends a simplified approach to control banding of nanomaterials (Figure 19). With this approach, nanomaterials are grouped into three hazard groups: (A) known to be inert, (B) understand reactivity and function, or (C) unknown properties. The exposure duration is described as Short (<4 hours/day, 2 days/week), Medium (4–6 hours/day, 3–5 days/week) or Long (>6 hours/day, 3–5 days/week). The potential for exposure is described through the state of the nanomaterial: bound (nanoparticles in a solid matrix), potential release (nanoparticles in friable matrix), or free/unbound (nanoparticles unbound, not aggregated). These elements are used to determine the recommended control band.

Exposure Duration	Bound Materials	Potential Release	Free / Unbound
Hazard Group A (Known to be inert)			
Shout	1	1	2
Medium	1	1	2
Long	1	2	2
Hazard Group B (Understand reactivity/function)			
Short	1	2	2
Medium	1	2	3
Long	1	3	3
Hazard Group C (Unknown Properties)			
Short	2	2	3
Medium	2	3	4
Long	2	4	4

- Band 1: Use good industrial hygiene practice and general ventilation.
- Band 2: Use an engineering control, typically local exhaust ventilation.
- Band 3: Enclose the process.
- Band 4: Seek expert advice.

Figure 19. GoodNanoGuide control banding matrix.

Another tool, the CB Nanotool, bases the control band for a particular task on the overall risk level (RL), which is determined by a "severity" score and a "probability" score (Figure 20). The severity score is determined by the sum of points assigned to the following factors: surface chemistry, particle shape, particle diameter, solubility, carcinogenicity, reproductive toxicity, mutagenicity, dermal toxicity, and hazard potential of the nanomaterial and the macro-parent material. The overall probability score is based on the following elements: estimated amount of nanomaterial used during the task, dustiness or mistiness, number of employees with similar exposures, frequency of operation, and duration of operation (Paik et al. 2008). The CB Nanotool is being used at the Lawrence Livermore National Laboratory (LLNL) and can be downloaded at http://controlbanding.net/Home.html.

One limitation of the CB Nanotool and other control banding tools for nanomaterials is that there are very few toxicological data on which to recommend control levels, other than the highest two levels, and to evaluate the validity of the tool. As health hazard studies continue to expand, and the understanding of the toxicity of nanomaterials improves, the severity scores may be adjusted to reflect the new knowledge and thereby affect the severity score to elicit a more defined control band (Zalk et al. 2009).

		Probability			
		Extremely Unlikely (0–25)	Less Likely (20–50)	Likely (51–75)	Probable (76–100)
Severity	Very High (76–100)	RL 3	RL 3	RL 4	RL 4
	High (51–75)	RL 2	RL 2	RL 3	RL 4
	Medium (26–50)	RL 1	RL 1	RL 2	RL 3
	Low (0–25)	RL 1	RL 1	RL 1	RL 2

Control bands:

RL 1: General ventilation
RL 2: Fume hoods or local exhaust ventilation
RL 3: Containment
RL 4: Seek specialist advice

Figure 20. Risk level matrix for the CB Nanotool.

The Australian Control Banding tool is specific to carbon nanotubes (Safe Work Australia 2012). The exposure potential is based on the amounts and types of activities, and determines the control band.

11. Fire and Explosion Control

Both carbon-containing and metal dusts can explode if they are aerosolized at a high enough concentration and if oxygen and an ignition source are present. Because the surface-to-volume ratio increases as a particle becomes smaller, nanoparticles may be more prone to explosion than an equivalent mass concentration of larger particles. In general, the potential and severity of nanomaterial explosions increase proportionally to the quantity of combustible nanomaterials being used. Thus, bench-scale research should present fewer explosion risks than work in pilot plants or full-scale manufacturing facilities. Nonetheless, all researchers should avoid creating large, highly concentrated aerosols of combustible nanomaterials.

12. Management of Nanomaterial Spills

Procedures should be developed to protect employees from exposure to nanomaterials during the cleanup of spills and spill-contaminated surfaces. Inhalation and dermal exposures will likely present the greatest risks. The potential for inhalation exposure during cleanup will be influenced by the likelihood of nanomaterials becoming airborne, with powder form presenting a greater inhalation potential than nanomaterials in solution, and liquids in turn presenting a greater potential risk than encapsulated nanomaterials.

Until relevant health and workplace exposure information is available, it is prudent to base strategies for dealing with spills and contaminated surfaces on the use of current good practices such as dust control and suppression. Standard approaches for cleaning powder spills can be used for cleaning surfaces contaminated with dry powder nanomaterials. These include access control, using HEPA-filtered vacuum cleaners, wiping up dry powders with damp cloths, or wetting the powder before wiping. Liquid spills containing nanomaterials can typically be cleaned by applying absorbent materials/liquid traps. If vacuum cleaning is employed, HEPA-filtered vacuums should be used, and care should be taken that HEPA filters are installed properly and that vacuum bags are changed according to the manufacturer's recommendations. Dry sweeping or air hoses should not be used to clean work areas. As in the case of any material spills or cleaning of contaminated surfaces, the handling and disposal should follow all applicable federal, state, and local regulations.

Equipment to contain and clean a nanomaterial spill should be readily available in or near each laboratory working with such materials. A nanomaterial spill kit for a laboratory environment may contain the following:

- Barricade tape.
- Nitrile or other chemically impervious gloves.
- Elastomeric respirator with appropriate filters.
- Adsorbent material.
- Wipes.
- Sealable plastic bags.
- Walk-off mat (e.g., Tacki-Mat®).

- HEPA-filtered vacuum.
- Spray bottle with deionized water or other appropriate liquid.

13. Occupational Health Surveillance

Occupational health surveillance involves the ongoing systematic col lection, analysis, and dissemination of exposure and health data on groups of workers for the purpose of preventing illness and injury (NIOSH 2009b). Occupational health surveillance, which includes hazard and medical surveillance, is an essential component of an effective occupational safety and health program (Harber et al. 2003; Baker and Matte 2005; NIOSH 2006; Wagner and Fine 2008). NIOSH continues to recommend occupational health surveillance as an important part of an effective risk management program for nanomaterial workers.

Medical screening in the workplace focuses on the early detection of health outcomes for individual workers and may involve an occupational history, medical examination, and application of specific medical tests to detect the presence of toxicants or early pathologic changes before the worker would normally seek clinical care for symptomatic presentations. Medical screening and resulting interventions represent secondary prevention and should not replace primary prevention efforts to minimize employee exposures to nano-materials. Medical surveillance involves the ongoing evaluation of the health status of a group of workers through the collection and aggregate analysis of health data for the purpose of preventing disease and evaluating the effectiveness of intervention programs.

Specific guidance for workers exposed to Carbon Nanotubes or Nanofibers is described in the Draft NIOSH *Current Intelligence Bulletin: Occupational Exposure to Carbon Nanotubes or Nanofibers* (NIOSH 2010a). NIOSH has developed interim guidance relevant to medical screening (one component of an occupational health surveillance program) for nanotechnology workers (see NIOSH *Current Intelligence Bulletin: Interim Guidance on Medical Screening of Workers Potentially Exposed to Engineered Nanoparticles* (http://www.cdc.gov/niosh/docs/2009-116)).

If medical screening recommendations exist for chemical or bulk materials of which nanomaterials are composed, they would apply to nanomaterials as well. A basic medical surveillance program should contain the following elements (Trout and Schulte 2010):

- An initial medical evaluation performed by a qualified health professional and other examinations or medical tests deemed necessary by the health professional.
- Periodic evaluations including symptoms surveys, physical exams, or specific medical tests based on data gathered in the initial evaluation.
- Post-incident evaluations.
- Employee training.
- Periodic analysis of the medical screening data to identify trends or patterns.

CONCLUSION

Given the growing body of knowledge about the potential hazards presented by worker exposure to engineered nanomaterials, it is important to protect researchers, laboratory staff, and others who work in the laboratory (e.g., janitors). The full range of occupational hygiene controls will be necessary to limit exposures to nanomaterials as a means to prevent adverse health outcomes in the research community. Engineering and administrative controls can eliminate or minimize the amount of nanomaterials that will be present in workplace air or settled on surfaces. Personal protective equipment can be used where other types of controls are not available or practical.

Nanomaterial health and safety is a rapidly evolving field that must respond to new information regarding nanomaterial toxicity and exposure potential. Thus, it is recommended that researchers and health and safety professionals stay abreast of new developments in nanomaterial workplace protection as they are published, both in the peer-reviewed literature and on credible Web sites such as those of NIOSH (www.cdc.gov/ niosh) and the GoodNanoGuide (www.goodnanoguide.org).

REFERENCES

ACGIH (2007). Industrial ventilation: A manual of recommended practice for design. 26th ed. Cincinnati, OH: American Congress of Governmental Industrial Hygienists.

Ahmad Z (2006). Processing and synthesis techniques for the preparation of nanomaterials. Dhahran, Saudi Arabia: Mechanical Engineering

Department, King Fahd University of Petroleum & Minerals (http://www.azonano.com/article.aspx?ArticleID=1710). Date accessed: July 20, 2011.

Aitken R, Creely KS, Tran CL (2004). Nanoparticles: an occupational hygiene review. London, UK: Health and Safety Executive (http://www.hse.gov.uk/research/rrhtm/rr274.htm). Date accessed: July 20, 2011.

Baker EL, Matte TP (2005). Occupational health surveillance. In: Rosenstock L, Cullen MR, Brodkin CA, Redlich CA, eds. Textbook of clinical occupational and environmental medicine. Philadelphia, PA: *Elsevier Saunders Company,* pp. 76–82.

Barlow PG, Clouter-Baker A, Donaldson K, Maccallum J, Stone V (2005). Carbon black nanoparticles induce type II epithelial cells to release chemotoxins for alveolar macrophages. *Part Fibre Toxicol 2*:11.

Behrens I, Pena AI, Alonso MJ, Kissel T (2002). Comparative uptake studies of bioadhesive and nonbioadhesive nanoparticles in human intestinal cell lines and rats: the effect of mucus on particle adsorption and transport. *Pharm Res 19*(8):1185–1193.

Brown DM, Wilson MR, MacNee W, Stone V, Donaldson K (2001). Size-dependent proinflammatory effects of ultrafine polystyrene particles: a role for surface area and oxidative stress in the enhanced activity of ultrafines. *Toxicol Appl Pharmacol 175*(3):191–199.

Burgess WA, Ellenbecker MJ, Treitman RD (2004). Ventilation for control of the work environment. 2nd ed. Hoboken, N.J.: *Wiley-Interscience,* pp. xvi.

Buzea C, Pacheco II, Robbie K (2007). Nanomaterials and nanoparticles: sources and toxicity. *Biointerphases 2*(4):MR17–MR71.

Chattopadhyay K (1992). Nanocomposites by rapid solidification route. *Bulletin of Materials Science 15*(6):515–525.

Chosewood LC, Wilson DE, eds. (2009). Biosafety in microbiological and biomedical laboratories. 5th ed. Washington, DC: U.S. Department of Health and Human Services, Public Health Service, Centers for Disease Control and Prevention, National Institutes of Health HHS Publication No.(CDC) 2–112.

Demou E, Stark WJ, Hellweg S (2009). Particle emission and exposure during nanoparticle synthesis in research laboratories. *Ann Occup Hyg 53*(8):829–838.

DiBerardinis LJ (1993). Guidelines for laboratory design: health and safety considerations. 2nd ed. New York: Wiley, pp. xiv.

Donaldson K, Aitken R, Tran L, Stone V, Duffin R, Forrest G, Alexander A (2006). Carbon nanotubes: a review of their properties in relation to pulmonary toxicology and workplace safety. *Toxicol Sci 92*(1):5–22.

Duffin R, Tran C, Clouter A, Brown D, MacNee W, Stone V, Donaldson K (2002). The importance of surface area and specific reactivity in the acute pulmonary inflammatory response to particles. *Ann Occup Hyg 46*(suppl. 1):242–245.

Duffin R, Tran L, Brown D, Stone V, Donaldson K (2007). Proinflammogenic effects of low-toxicity and metal nanoparticles in vivo and in vitro: highlighting the role of particle surface area and surface reactivity. *Inhalation Toxicol 19*(10):849–856.

Elder A, Gelein R, Silva V, Feikert T, Opanashuk L, Carter J, Potter R, Maynard A, Ito Y, Finkelstein J, Oberdörster G (2006). Translocation of inhaled ultrafine manganese oxide particles to the central nervous system. *Environ Health Perspect 114*(8):1172–1178.

Erdely A, Hulderman T, Salmen R, Liston A, Zeidler-Erdely PC, Schwegler-Berry D, Castranova V, Koyama S, Kim Y-A, Endo M, Simeonova PP (2009). Cross-talk between lung and systemic circulation during carbon nanotube respiratory exposure. *Potential Biomarkers. Nano Letters 9*(1):36–43.

Geiser M, Rothen-Rutishauser B, Kapp N, Schürch S, Kreyling W, Schulz H, Semmler M, Im Hof V, Hey-der J, Gehr P (2005). Ultrafine particles cross cellular membranes by nonphagocytic mechanisms in lungs and in cultured cells. *Environ Health Perspect 113*(11):1555–1560.

Göhler D, Stintz M, Hillemann L, Vorbau M (2010). Characterization of nanoparticle release from surface coatings by the simulation of a sanding process. *Ann Occup Hyg 54*(6):615–624.

Harber P, Muranko H, Shvartsblat S, Solis S, Torossian A, Oren T (2003). A triangulation approach to historical exposure assessment for the carbon black industry. *J Occup Environ Med 45*(2):131–143.

Harford AJ, Edwards JW, Priestly BG, Wright PFA (2007). Current OHS best practices for the Austrialian nanotechnology industry. A position paper by the NanoSafe Australia Network. Melborne, Australia: The NanoSafe Australia Network.

He CN, Zhao NQ, Shi CS, Song SZ (2009). Fabrication of carbon nanomaterials by chemical vapor deposition. *Journal of Alloys and Compounds 484*(1–2):6–11.

Holoman MN, Novotny M, Kemsley M, Raje J, Sullivan J, Pekarskaya T, et al. (2007). The nanotech report. New York: Lux Research.

Huang RF, Wu YD, Chen HD, Chen C-C, Chen C-W, Chang C-P, Shih T-S (2007). Development and evaluation of an air-curtain fume cabinet with considerations of its aerodynamics. *Ann Occup Hyg 51*(2):189–206.

ICRP (1994). Human respiratory tract model for radiologic protection. ICRP Publication 66. Ottawa, Ontario, Canada: International Commission on Radiological Protection (http://www.elsevier.com/ wps/find/book description.cws_home/29164/description#description). Date accessed: July 20, 2011.

Johnson AE, Fletcher B (1996). The effect of operating conditions on fume cupboard containment. *Safety Science 24*(1):51–60.

Johnson DR, Methner MM, Kennedy AJ, Steevens JA (2010). Potential for occupational exposure to engineered carbon-based nanomaterials in environmental laboratory studies. *Environ Health Perspect 118*(1):49–54.

Ju S-Y, Kopcha WP, Papadimitrakopoulos F (2009). Brightly fluorescent single-walled carbon nanotubes via an oxygen-excluding surfactant organization. *Science 323*(5919):1319–1323.

Kane AB, Hurt RH (2008). Nanotoxicology: the asbestos analogy revisited. Nat Nano *3*(7):378–379, doi:10.1038/nnano.2008.182.

Karlsson HL, Gustafsson J, Cronholm P, Möller L (2009). Size-dependent toxicity of metal oxide particles—a comparison between nano- and micrometer size. *Toxicol Lett 188*(2):112–118.

Kisin ER, Murray AR, Keane MJ, Shi XC, Schwegler-Berry D, Gorelik O, Arepalli S, Castranova V, Wallace WE, Kagan VE, Shvedova AA (2007). Single-walled carbon nanotubes: Geno- and cytotoxic effects in lung fibroblast V79 cells. *J Toxicol Environ Health Part A 70*(24):2071–2079.

Lam CW, James JT, McCluskey R, Arepalli S, Hunter RL (2006). A review of carbon nanotube toxicity and assessment of potential occupational and environmental health risks. Crit Rev Toxicol *36*(3):189–217. Lam CW, James JT, McCluskey R, Hunter RL (2004). Pulmonary toxicity of single-wall carbon nanotubes in mice 7 and 90 days after intratracheal instillation. *Toxicol Sci 77*(1):126–34.

Li Z, Hulderman T, Salmen R, Chapman R, Leonard SS, Young SH, Shvedova A, Luster MI, Simeonova PP (2007). Cardiovascular effects of pulmonary exposure to single-wall carbon nanotubes. *Environ Health Perspect 115*(3):377–382.

Lison D, Lardot C, Huaux F, Zanetti G, Fubini B (1997). Influence of particle surface area on the toxicity of insoluble manganese dioxide dusts. *Arch Toxicol 71*(12):725–729.

Ma-Hock L, Treumann S, Strauss V, Brill S, Luizi F, Mertler M, Wiench K, Gamer AO, van Ravenzwaay B, Landsiedel R (2009). Inhalation toxicity of multiwall carbon nanotubes in rats exposed for 3 months. *Toxicol Sci 112*(2):468–481.

Maynard AD, Kuempel ED (2005). Airborne nanostructured particles and occupational health. *J Nanopart Res 7*(6):587–614.

Miyawaki J, Yudasaka M, Azami T, Kubo Y, Iijima S (2008). Toxicity of single-walled carbon nanohorns. *ACS Nano 2*(2):213–226.

Murashov V, Howard J (2009). Essential features for proactive risk management. *Nat Nanotechnol 4*(8):467–470.

Murray AK, Kisin E, Kommineni C, Kagen VE, Castranova V, Shvedova AA (2007). Single-walled carbon nanotubes induce oxidative stress and inflammation in skin. *Toxicologist 96*(1)*:*A1406.

Nemmar A, Hoet PHM, Thomeer M, Nemery B, Vanquickenborne B, Vanbilloen H, Mortelmans L, Hoylaerts MF, Verbruggen A, Dinsdale A (2002). Passage of inhaled particles into the blood circulation in humans—response. *Circulation 106*(20):E141–E142.

NIOSH (1994). NIOSH manual of analytical methods (NMAM®), Diesel particulate matter (supplement issues 3/15/03). 4th ed. By Schlecht PC, O'Conner PF, eds. Cincinnati, OH: U.S. Department of Health and Human Services, Centers for Disease Control and Prevention, National Institute for Occupational Safety and Health. DHHS (NIOSH) Publication No. 9–13 (www.cdc.gov/niosh/nmam/).

NIOSH (2005). NIOSH respirator selection logic. Cincinnati, OH: U.S. Department of Health and Human Services, Centers for Disease Control and Prevention, National Institute for Occupational Safety and Health. DHHS (NIOSH) Publication No. 200–00 (http://www.cdc.gov/niosh /docs/2005-100/).

NIOSH (2006). Criteria for a recommended standard: occupational exposure to refractory ceramic fibers. Cincinnati, OH: U.S. Department of Health and Human Services, Centers for Disease Control and Prevention, National Institute for Occupational Safety and Health. DHHS (NIOSH) Publication No. 200–23 (www.cdc.gov/niosh/docs/2006-123).

NIOSH (2007). Progress toward safe nanotechnology in the workplace: a report from the NIOSH nano-technology research center. Cincinnati, OH: U.S. Department of Health and Human Services, Centers for Disease Control and Prevention, National Institute for Occupational Safety and Health. DHHS (NIOSH) Publication No. 200–23 (www.cdc.gov/niosh /docs/2007-123).

NIOSH (2008). Safe nanotechnology in the workplace. Cincinnati, OH: U.S. Department of Health and Human Services, Centers for Disease Control and Prevention, National Institute for Occupational Safety and Health. DHHS (NIOSH) Publication No. 200–12 (www.cdc.gov/niosh/docs/2008-112).

NIOSH (2009a). Approaches to safe nanotechnology: managing the health and safety concerns associated with engineered nanomaterials. Cincinnati, OH: U.S. Department of Health and Human Services, Centers for Disease Control and Prevention, National Institute for Occupational Safety and Health. DHHS (NIOSH) Publication No. 200–25 (www.cdc.gov/niosh /docs/2009-125/pdfs/2009-125.pdf).

NIOSH (2009b). Interim guidance for medical screening and hazard surveillance for workers potentially exposed to engineered nanoparticles. Cincinnati, OH: U.S. Department of Health and Human Services, Centers for Disease Control and Prevention, National Institute for Occupational Safety and Health. DHHS (NIOSH) Publication No. 200–16 (www.cdc.gov/niosh/docs/2009-116).

NIOSH (2009c). Qualitative risk characterization and management of occupational hazards: control banding (CB). Cincinnati, OH: U.S. Department of Health and Human Services, Centers for Disease Control and Prevention, National Institute for Occupational Safety and Health. DHHS (NIOSH) Publication No. 200–52 (http://www.cdc.gov/niosh /docs/2009-152/).

NIOSH (2010a). (Draft) NIOSH Current intelligence bulletin. Occupational exposure to carbon nano-tubes and carbon nanofibers. Cincinnati, OH: U.S. Department of Health and Human Services, Centers for Disease Control and Prevention, National Institute for Occupational Safety and Health (http:www.cdc.gov/niosh/docket/review/docket 161A/pdfs/carbon NanotubeCIB_PublicReviewOfDraft.pdf). Date accessed: July 20, 2011.

NIOSH (2010b). Prevention through Design. Cincinnati, OH: U.S. Department of Health and Human Services, Centers for Disease Control and Prevention, National Institute for Occupational Safety and Health (http://www.cdc.gov/niosh/topics/ptd/). Date accessed: October 17, 2010.

NIOSH (2011). Evaluation of health hazard and recommendations for occupational exposure to titanium dioxide. Cincinnati, OH: U.S. Department of Health and Human Services, Centers for Disease Control and Prevention, National Institute for Occupational Safety and Health. NIOSH Publication No. 201–60 (www.cdc.gov/niosh/docs/2011-160/). Date accessed: August 29, 2011.

NRC (2011). Prudent practices in the laboratory: handling and management of chemical hazards (updated version). Washington, DC: National Research Council of the National Academies.

NSTC (2007). The national nanotechnology initiative strategic plan. Washington, DC: National Science and Technology Council (http://www.nano.gov/NNI_Strategic_Plan_2007.pdf). Date accessed: July 20, 2011.

Oberdörster G, Ferin J, Gelein R, Soderholm SC, Finkelstein J (1992). Role of the alveolar macrophage in lung injury: studies with ultrafine particles. *Environ Health Perspect 97*:193–199.

Oberdörster G, Ferin J, Lehnert BE (1994). Correlation between particle size, in vivo particle persistence, and lung injury. *Environ Health Perspect 102*(Suppl 5):173–179.

Oberdörster G, Maynard A, Donaldson K, Castranova V, Fitzpatrick J, Ausman K, Carter J, Karn B, Kreyling W, Lai D, Olin S, Monteiro-Riviere N, Warheit D, Yang H (2005). Principles for characterizing the potential human health effects from exposure to nanomaterials: elements of a screening strategy. *Part Fibre Toxicol 2*:8.

Oberdörster G, Sharp Z, Atudorei V, Elder A, Gelein R, Kreyling W, Cox C (2004). Translocation of inhaled ultrafine particles to the brain. *Inhal Toxicol 16*(6–7):437–445.

Oberdörster G, Sharp Z, Atudorei V, Elder A, Gelein R, Lunts A, Kreyling W, Cox C (2002). Extrapulmonary translocation of ultrafine carbon particles following whole-body inhalation exposure of rats. *J Toxicol Environ Health A 65*(20):1531–1543.

OECD (2002). Frascati manual, 2002: proposed standard practice for surveys on research and experimental development, the measurement of scientific and technological activities. 6th ed. Paris: Organisation For Economic Co-operation and Development, p. 255.

OSHA (1990).29 CFR 1910.1450 Occupational exposure to hazardous chemicals in laboratories (www. osha.gov). Washington, DC: Occupational Safety and Health Administration. Date accessed: July 20, 2011.

OSHA (1992). General industry, safety and health standards. Washington, DC: Occupational Safety and Health Administration (http://www.osha.gov/pls /oshaweb/owadisp.show_document?p_id=12716&p_table=standards). Date accessed: July 20, 2011.

OSHA (1994). 29 CFR 1910.1200. Hazard communication, toxic and hazardous substances. Washington, DC: Occupational Safety and Health Administration. Date accessed: December 15, 2011.

OSHA (2008). 29 CFR 1910.132. General requirements: personal protective equipment. Washington, DC: Occupational Safety and Health Administration. Date accessed: December 15, 2011.

Paik SY, Zalk DM, Swuste P (2008). Application of a pilot control banding tool for risk level assessment and control of nanoparticle exposures. *Ann Occup Hyg 52*(6):419–428.

Pauluhn J (2010). Subchronic 13-week inhalation exposure of rats to multiwalled carbon nanotubes: toxic effects are determined by density of agglomerate structures, not fibrillar structures. *Toxicol Sci 113*(1):226–242.

Poland CA, Duffin R, Kinloch I, Maynard A, Wallace WAH, Seaton A, Stone V, Brown S, MacNee W, Donaldson K (2008). Carbon nanotubes introduced into the abdominal cavity of mice show asbestos-like pathogenicity in a pilot study. *Nat Nanotechnol 3*(7):423–428.

Rao CN, Biswas K (2009). Characterization of nanomaterials by physical methods. *Annu Rev Anal Chem 2*:435–462.

Ryman-Rasmussen JP, Riviere JE, Monteiro-Riviere NA (2006). Penetration of intact skin by quantum dots with diverse physicochemical properties. *Toxicol Sci 91*(1):159–165.

Safe Work Australia (2012). Safe handling and use of carbon nanotubes. Commonwealth Scientific and Industrial Research Organisation (CSIRO) and Safe Work Australia. ISBN 978–0–642–33351–3.

Sahu M, Biswas P (2010). Size distributions of aerosols in an indoor environment with engineered nanoparticle synthesis reactors operating under different scenarios. *J Nanopart Res 12*(3):1055–1064.

Samuneva B, Djambaski P, Kashchieva E, Chernev G, Salvado IMM, Fernandes MHV, Wu A (2006). Sol gel synthesis and structure of hybrid nanomaterials with strong chemical bonds. In: Gdoutos EE, ed. Fracture of nano and engineering materials and structures. Dordrecht, The Netherlands: *Springer,* pp. 1037–1038.

Saunders GT (1993). Laboratory fume hoods: a users manual. New York: *John Wiley & Sons,* Inc.

Schulte HF, Hyatt EC, Jordan HS, Mitchell RN (1954). Evaluation of laboratory fume hoods. *Am Ind Hyg Assoc Q 15*(3):195–202.

Schulte P, Geraci C, Zumwalde R, Hoover M, Kuempel E (2008a). Occupational risk management of engineered nanoparticles. *J Occup Environ Hyg 5*(4):239–249.

Schulte P, Rinehart R, Okun A, Geraci C, Heidel D (2008b). National Prevention through Design Initiative. *J Safety Research 39*:115–121.

Seaton A, Tran L, Aitken R, Donaldson K (2010). Nanoparticles, human health hazard and regulation. *J R Soc Interface 7*(Suppl 1):S119–129.

Shvedova AA, Castranova V, Kisin ER, Schwegler-Berry D, Murray AR, Gandelsman VZ, Maynard A, Baron P (2003). Exposure to carbon nanotube material: assessment of nanotube cytotoxicity using human keratinocyte cells. *J Toxicol Environ Health* A *66*(20):1909_1926.

Shvedova AA, Kisin E, Murray AR, Johnson VJ, Gorelik O, Arepalli S, Hubbs AF, Mercer RR, Keohavong P, Sussman N, Jin J, Yin J, Stone S, Chen BT, Deye G, Maynard A, Castranova V, Baron PA, Kagan VE (2008). Inhalation vs. aspiration of single-walled carbon nanotubes in C57BL/6 mice: inflammation, fibrosis, oxidative stress, and mutagenesis. *Am J Physiol Lung Cell Mol Physiol 295*(4):L552–L565.

Shvedova AA, Kisin ER, Mercer R, Murray AR, Johnson VJ, Potapovich AI, Tyurina YY, Gorelik O, Arepalli S, Schwegler-Berry D, Hubbs AF, Antonini J, Evans DE, Ku B-K, Ramsey D, Maynard A, Kagan VE, Castranova V, Baron P (2005). Unusual inflammatory and fibrogenic pulmonary responses to single-walled carbon nanotubes in mice. *Am J Physiol Lung Cell Mol Physiol 289*(5):L698–L708.

Shvedova AA, Kisin ER, Porter D, Schulte P, Kagan VE, Fadeel B, Castranova V (2009). Mechanisms of pulmonary toxicity and medical applications of carbon nanotubes: two faces of Janus? *Pharmacol Ther 121*(2):192–204.

Singh N, Manshian B, Jenkins GJ, Griffiths SM, Williams PM, Maffeis TG, Wright CJ, Doak SH (2009). NanoGenotoxicology: the DNA damaging potential of engineered nanomaterials. *Biomaterials 30*(23–24):3891–3914.

Takagi A, Hirose A, Nishimura T, Fukumori N, Ogata A, Ohashi N, Kitajima S, Kanno J (2008). Induction of mesothelioma in p53+/- mouse by intraperitoneal application of multi-wall carbon nanotube. *J Toxicol Sci 33*(1):105–116.

Takenaka S, Karg E, Roth C, Schulz H, Ziesenis A, Heinzmann U, Schramel P, Heyder J (2001). Pulmonary and systemic distribution of inhaled ultrafine silver particles in rats. *Environ Health Perspect 1099*(Suppl 4):547–551.

Tinkle SS, Antonini JM, Rich BA, Roberts JR, Salmen R, DePree K, Adkins EJ (2003). Skin as a route of exposure and sensitization in chronic beryllium disease. *Environ Health Perspect 111*(9):1202–1208.

Tran C, Cullen R, Buchanan D, Jones A, Miller B, Searl A, Davis J, Donaldson K (1999). Investigation and prediction of pulmonary responses to dust, part II. In: Investigation into the pulmonary effects of low toxicity dusts. (No. 216/1999). Suffolk, UK: Health and Safety Executive.

Tran CL, Buchanan D, Cullen RT, Searl A, Jones AD, Donaldson K (2000). Inhalation of poorly soluble particles. II. Influence of particle surface area on inflammation and clearance. *Inhal Toxicol 12*(12):1113–1126.

Trout DB, Schulte PA (2010). Medical surveillance, exposure registries, and epidemiologic research for workers exposed to nanomaterials. *Toxicology 269*(2–3):128–135.

Tsai S-J, Ada E, Isaacs J, Ellenbecker M (2009a). Airborne nanoparticle exposures associated with the manual handling of nanoalumina and nanosilver in fume hoods. *J Nanopart Res 11*(1):147–161.

Tsai S-J, Hofmann M, Hallock M, Ada E, Kong J, Ellenbecker M (2009b). Characterization and evaluation of nanoparticle release during the synthesis of single-walled and multiwalled carbon nanotubes by chemical vapor deposition. *Environ Sci Technol 43*(15):6017–6023.

Tsai S-J, Huang RF, Ellenbecker MJ (2010). Airborne nanoparticle exposures while using constant-flow, constant-velocity, and air-curtain-isolated fume hoods. *Ann Occup Hyg 54*(1):78–87.

UNH (2009). Nanomaterials safety program. Durham, NH: University of New Hampshire (http://www. unh.edu/research/sites/unh.edu.research/files /docs/EHS/Chem-safety/UNH-Nanomaterials-SafetyProgram.pdf). Date accessed: July 20, 2011.

UNC (2011). Summary of recommended nanomaterial risk levels (NRL). Chapel Hill, NC: University of North Carolina at Chapel Hill (http://ehs.unc.edu/ih/lab/nano.shtml). Date accessed: July 20, 2011.

Wagner GF, Fine LJ (2008). Surveillance and health screening in occupational health. In: Wallace R, ed., Maxcy-Rosenau-Last. Public health and preventive medicine. 15th ed. New York: McGraw-Hill Medical Publishing, pp. 759–793.

Zalk D, Paik S, Swuste P (2009). Evaluating the control banding nanotool: a qualitative risk assessment method for controlling nanoparticle exposures. *J Nanopart Res 11*(7):1685–1704, doi: 10.1007/ s11051-009-9678-y.

INDEX

D

E

F

G

H

I

J

K

L

M

N

R

S

T

U

V

W

X

Y